3|21

RACKET PROGRAMMING THE FUN WAY

RACKET PROGRAMMING THE FUN WAY

From Strings to Turing Machines

by James W. Stelly

no starch press

San Francisco

Printed in the United States of America

First printing

25 24 23 22 21 1 2 3 4 5 6 7 8 9

ISBN-13: 978-1-7185-0082-2 (print)
ISBN-13: 978-1-7185-0083-9 (ebook)

Publisher: William Pollock
Executive Editor: Barbara Yien
Production Editor: Dapinder Dosanjh
Developmental Editor: Alex Freed
Interior Design: Octopod Studios
Cover Illustration: Gina Redman
Technical Reviewer: Matthew Flatt
Copyeditor: Chris Cartwright
Proofreader: Emelie Battaglia

For information on distribution, translations, or bulk sales, please contact No Starch Press, Inc. directly:
No Starch Press, Inc.
245 8th Street, San Francisco, CA 94103
phone: 415.863.9900; fax: 415.863.9950; info@nostarch.com
www.nostarch.com

Library of Congress Cataloging-in-Publication Data
Names: Stelly, James W., author.
Title: Racket programming the fun way: from strings to turing machines / by James W. Stelly.
Description: San Francisco : No Starch Press, [2021]. | Includes
bibliographical references and index.
Identifiers: LCCN 2020022884 (print) | LCCN 2020022885 (ebook) | ISBN
9781718500822 | ISBN 9781718500839 (ebook) | ISBN 1718500822
Subjects: LCSH: Racket (Computer program language) | LISP (Computer program
language) | Computer programming.
Classification: LCC QA76.73.R33 S 2020 (print) | LCC QA76.73.R33 (ebook)
| DDC 005.13/3-dc23
LC record available at https://lccn.loc.gov/2020022884
LC ebook record available at https://lccn.loc.gov/2020022885

I dedicate this book to my mom and dad who patiently (and at times not so patiently) endured my many childhood pranks.

About the Author

James W. Stelly has been dabbling with computers as both a hobbyist and professional for over four decades. He has degrees in both computer science and mathematics from the University of Houston. As a hobbyist, his projects include robotics using Arduino and Raspberry Pi along with numerous explorations of programming languages ranging from machine language to C++ (and many others). His day job (now part time since retirement) is developing a line of business applications primarily aimed at record keeping and data management.

About the Technical Reviewer

Matthew Flatt is a professor in the School of Computing at the University of Utah, where he works on extensible programming languages, run-time systems, and applications of functional programming. He is one of the developers of the Racket programming language and a co-author of the introductory programming textbook *How to Design Programs*.

BRIEF CONTENTS

CONTENTS IN DETAIL

ACKNOWLEDGMENTS

First, let me thank the folks responsible for Racket. They have created a truly remarkable piece of software. The care and effort that went into producing it has to be incalculable.

I would like to extend my deepest gratitude and thanks to my editors at No Starch, Alex Freed and Athabasca Witschi, as well as my technical reviewer, Matthew Flatt. They made literally dozens of helpful suggestions for improvements as well as corrections. Any remaining flaws are entirely of my own making.

Finally, let me thank my wife for her patience. Her view of me was primarily of the back of my head while I was working on this book.

INTRODUCTION

In this book we explore using Racket (a language descended from the Scheme family of programming languages—which in turn descended from Lisp) and DrRacket, a graphical environment that allows us to make the most of all the features of Racket. One of the attractive features of this ecosystem is that it's equipped with a plethora of libraries that cover a wide range of disciplines. The developers describe Racket as a system that has "batteries included." This makes it an ideal platform for the interactive investigation of various topics in computer science and mathematics.

Given Racket's Lisp pedigree, we would be remiss to omit functional programming, so we will definitely explore it in this text. Racket is no one-trick pony though, so we will also explore imperative, object oriented, and logic programming along the way. Also on the computer science front, we will look at various abstract computing machines, data structures, and a number of search algorithms as related to solving some problems in recreational mathematics. We will finish the book by building our own calculator, which will entail lexical analysis using regular expressions, defining the grammar using extended Backus–Naur form (EBNF), and building a recursive descent parser.

Racket

Racket features extensive and well-written documentation, which includes *Quick: An Introduction to Racket with Pictures*, the introductory *Racket Guide*, and the thorough *Racket Reference*. Various other toolkits and environments also have separate documentation. Within DrRacket these items can be accessed through the Help menu.

Racket is available for a wide variety of platforms: Windows, Linux, macOS, and Unix. It can be downloaded from the Racket website via the link *https://download.racket-lang.org/*. Once downloaded, installation simply entails running the downloaded executable on Windows, *.dmg* file on macOS, or shell script on Linux. At the time of writing, the current version is 7.8. Examples in the book will run on any version 7.0 or later. They will likely run on earlier versions as well, but since the current version is freely available there is really no need to do so. When the DrRacket environment is first launched, the user will be prompted to select a Racket language variant. The examples in this book all use the first option in the pop-up dialog box (that is, the one that says "The Racket Language").

The DrRacket window provides a definitions pane (top pane in Figure 1) where variables and functions are defined and an interactions pane (bottom pane in Figure 1) where Racket code can be interactively executed. Within these panes, help is a single keypress away. Just click on any built-in function name and press F1.

The definitions window contains all the features one expects from a robust interactive development environment (IDE) such as syntax highlighting, variable renaming, and an integrated debugger.

Racket enthusiasts are affectionately known as *Racketeers* (catchy, eh?). Once you've had an opportunity to explore this wonderful environment, don't be surprised if you become a Racketeer yourself.

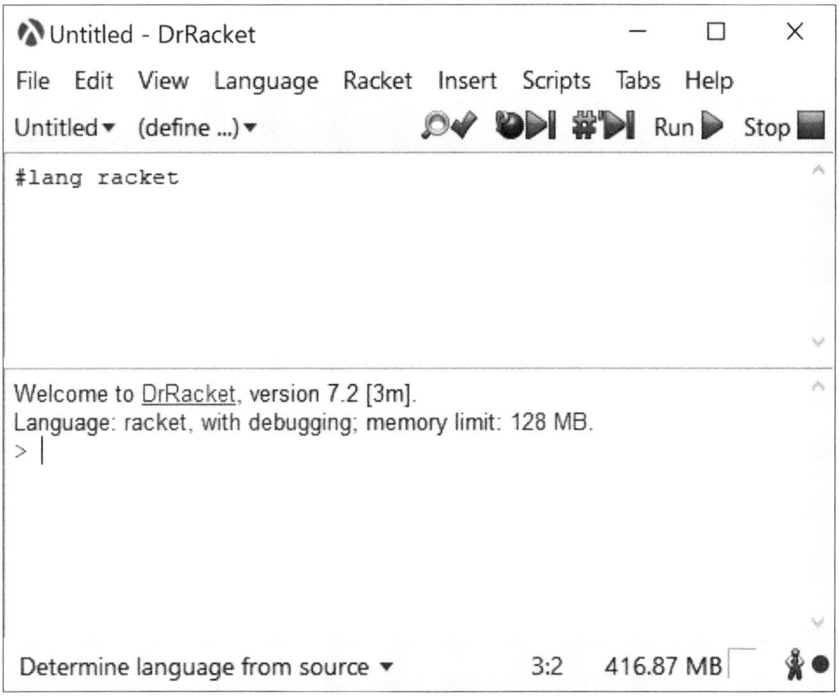

Figure 1: DrRacket IDE

Conventions Used in This Book

DrRacket supports a number of programming and learning languages. In this book we focus exclusively on the default Racket language. Thus, unless otherwise stated, all definition files should begin with the line

```
#lang racket
```

Code entered in the definitions section will be shown in a framed box as above.

Expressions entered in the interactive pane will be shown prefixed with a right angle bracket > as shown below. The angle bracket is DrRacket's input prompt. Outputs will be shown without the angle bracket. To easily differentiate inputs and outputs, inputs will be shown in bold in this book (but they are not bold in the IDE).

```
> (+ 1 2 3) ; this is an input, the following is an output
6
```

We occasionally make use of some special symbols that DrRacket supports, such as the Greek alphabet (for example, we may use θ as an identifier for an angle). These symbols are listed in Appendix B. The method used to enter these symbols is also given there. If you're typing the examples in by hand and don't want to use the special symbols, simply substitute a name of your choosing: for example use `alpha` instead of α.

An example of a program listing entered in the definitions window is shown below.

```
#lang racket

(define (piscis x y r b)
  (let* ([y (- y r)]
         [2r (* 2 r)]
         [yi (sqrt (- (sqr r) (sqr x)))] ; y-intersection
         [π pi]
       ❶ [ϕ (asin (/ yi r))]
       ❷ [θ (- π ϕ)]
       ❸ [path (new dc-path%)])
    (send dc set-brush b)
  ❹ (send path move-to 0 (- yi))
  ❺ (send path arc (- x r)     y 2r 2r  θ     (+ π  ϕ))
  ❻ (send path arc (- (- x) r) y 2r 2r (- ϕ) ϕ)
  ❼ (send dc draw-path path)))
```

We'll use Wingdings symbols such as ❶ to highlight interesting portions of the code.

Who This Book Is For

While no prior knowledge of Racket, Lisp, or Scheme is required, it wouldn't hurt to have some basic programming knowledge, but this is certainly not required. The mathematical prerequisites will vary. Some topics may be a bit challenging, but nothing more than high school algebra and trigonometry is assumed. A theorem or two may surface, but the treatment will be informal.

About This Book

If you're already familiar with the Racket language, feel free to skip (or perhaps just skim) the first couple of chapters as these just provide an introduction to the language. These early chapters are by no means a comprehensive encyclopedia of Racket functionality. The ambitious reader should consult the excellent Racket Documentation for fuller details. Here is a brief description of each chapter's content.

Chapter 1: Racket Basics Gives the novice Racket user a grounding in some of the basic Racket concepts that will be needed to progress through the rest of the book.

Chapter 2: Arithmetic and Other Numerical Paraphernalia Describes Racket's extensive set of numeric data types: integers, true rational numbers, and complex numbers (to name a few). This chapter will make the reader adept at using these entities in Racket.

Chapter 3: Function Fundamentals Introduces Racket's multi-paradigm programming capability. This chapter introduces the reader to both functional and imperative programming. The final section will look at a few fun programming applications.

Chapter 4: Plotting, Drawing, and a Bit of Set Theory Introduces interactive graphics. Most IDEs are textual only; DrRacket has extensive capability for generating graphical output in an interactive environment. This chapter will show you how it's done.

Chapter 5: GUI: Getting Users Interested Shows how to construct mini graphics applications that run in their own window.

Chapter 6: Data Explores various ways of handling data in Racket. It will discuss how to read and write data to and from files on your computer. It will also discuss ways to analyze data using statistics and data visualization.

Chapter 7: Searching for Answers Examines a number of powerful search algorithms. These algorithms will be used to solve various problems and puzzles in recreational mathematics.

Chapter 8: Logic Programming Takes a look at another powerful programming paradigm. Here we explore using Racket's Prolog-like logic programming library: Racklog.

Chapter 9: Computing Machines Takes a quick look at various abstract computing machines. These simple mechanisms are a gateway into some fairly deep concepts in computer science.

Chapter 10: TRAC: The Racket Algebraic Calculator Leverages skills developed in the previous chapters to build a stand-alone interactive command line calculator.

1

RACKET BASICS

Let's begin with an introduction to some basic concepts in Racket. In this chapter, we'll cover some of the fundamental data types that will be used throughout the book. You'll want to pay particular attention to the discussion of lists, which underpin much of Racket's functionality. We'll also cover how to assign values to variables and various ways to manipulate strings, and along the way, you'll encounter a first look at vectors and structs. The chapter wraps up with a discussion on how to produce formatted output.

Atomic Data

Atomic data is the basic building block of any programming language, and Racket is no exception. Atomic data refers to elementary data types that are typically considered to be indivisible entities; that is, numbers like `123`, strings like `"hello there"`, and identifiers such as `pi`. Numbers and strings evaluate to themselves; if bound, identifiers evaluate to their associated value:

```
> 123
123

> "hello there"
"hello there"

> pi
3.141592653589793
```

Evaluating an unbound identifier results in an error. To prevent an unbound identifier from being evaluated, you can prefix it with an apostrophe:

```
> alpha
. . alpha: undefined;
  cannot reference an identifier before its definition

> 'alpha
'alpha
```

We can organize atomic data together using lists, which are covered next.

Lists

In Racket, lists are the primary non-atomic data structures (that is, something other than a number, string, and so on). Racket relies heavily on lists because it's a descendant of *Lisp* (short for LISt Processing). Before we get into the details, let's look at some simple representative samples.

A First Look at Lists

Here's how to make a list with some numbers:

```
> (list 1 2 3)
```

Notice the syntax. Lists typically begin with an open parenthesis, (, followed by a list of space-separated items and end with a closed parenthesis,). The first item in the list is normally an identifier that indicates how the list is to be evaluated.

Lists can also contain other lists.

```
> (list 1 (list "two" "three") 4 5)
```

which prints as

```
'(1 ("two" "three") 4 5)
```

Note the apostrophe (or tick mark) at the beginning of the last example. This is an alias for the quote keyword. If you want to enter a literal list (a list that is simply accepted as is), you can enter it *quoted*:

```
> (quote (1 ("two" "three") 4 5))
```

or

```
> '(1 ("two" "three") 4 5)
```

Either of which print as

```
'(1 ("two" "three") 4 5)
```

While list and quote seem like two equivalent ways to build lists, there's an important difference between them. The following sequence illustrates the difference.

```
> (quote (3 1 4 pi))
'(3 1 4 pi)

> (list 3 1 4 pi)
'(3 1 4 3.141592653589793)
```

Notice that quote returns the list exactly as it was entered, but when list was used, the identifier pi was evaluated and its value was substituted in its place. In general, in a non-quoted list, *all* identifiers are evaluated and replaced by their associated values. The keyword quote plays an important role in macros and symbolic expression evaluation, which are advanced topics that we will not cover in this text.

One criticism of the Lisp family of languages is the proliferation of parentheses. To alleviate this, Racket allows either square brackets or curly brackets to be used instead. For example, it's perfectly acceptable to write the last expression as

```
> '(1 ["two" "three"] 4 5)
```

or

```
> '(1 {"two" "three"} 4 5)
```

S-Expressions

A list is a special case of something called an *s-expression*. An s-expression (or symbolic expression) is defined as being one of two cases:

Case 1 The s-expression is an atom.

Case 2 The s-expression is expression of the form (x . y) where *x* and *y* are other s-expressions.

The form `(x . y)` is typically called a *pair*. This is a special syntactic form used to designate a *cons* cell, which we will have much more to say about shortly.

Let's see if we can construct a few examples of s-expressions. Ah, how about `1`? Yes, it's an atom, so it satisfies case 1. What about `"spud"`? Yep, strings are atoms, and thus `"spud"` is also an s-expression. We can combine these to make another s-expression: `(1 . "spud")`, which satisfies case 2. Since `(1 . "spud")` is an s-expression, case 2 allows us to form another s-expression as `((1 . "spud") . (1 . "spud"))`. We can see from this that s-expressions are actually tree-like structures as illustrated in Figure 1-1. (Technically s-expressions form a *binary tree*, where non-leaf nodes have exactly two child nodes).

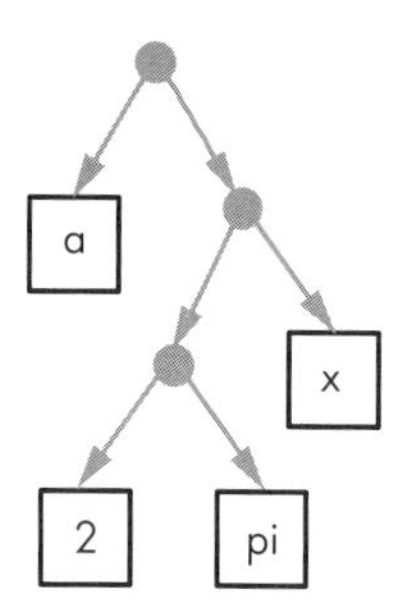

Figure 1-1: `((a . (2 . pi) . x))`

In Figure 1-1, the square boxes are leaf nodes representing atoms, and the circle nodes represent pairs. We'll see how s-expressions are used to construct lists in the next section.

List Structure

As mentioned above, a list is a special case of an s-expression. The difference is that, in a list, if we follow the rightmost elements in each pair, the final node is a special atomic node called *nil*. Figure 1-2 illustrates what the list `'(1 2 3)`—which as an s-expression is `(1 . (2 . (3 . nil)))`—looks like internally.

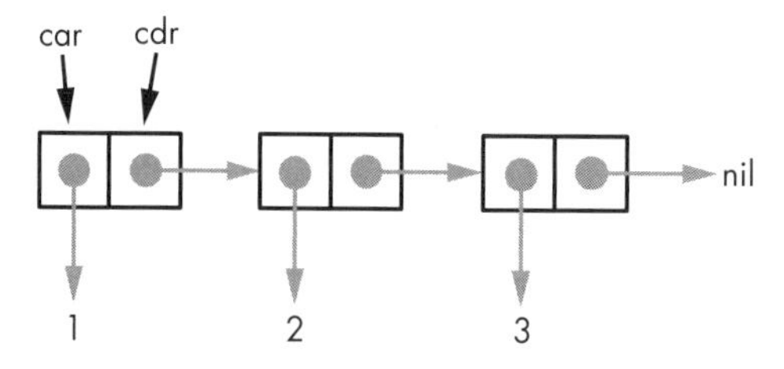

Figure 1-2: List structure

We've flattened the tree to better resemble a list. We've also expanded each pair node (aka a *cons cell*) to show that it consists of two cells, each of which contains a pointer to another node. These pointer cells, for historical reasons, are called *car* and *cdr* respectively (the names of computer registers used in early versions of Lisp). We can see that the last cdr cell in the list is pointing to nil. Nil is indicated in Racket by an empty list: `'()` or `null`.

Cons cells can be created directly by using the `cons` function. Note that the `cons` function does not necessarily create a list. For example

```
> (cons 1 2)
'(1 . 2)
```

produces a pair but *not* a list. However, if we use an empty list as our second s-expression

```
> (cons 1 '())
'(1)
```

we produce a list with just one element.

Racket provides a couple of functions to test whether something is a list or a pair. Note in Racket `#t` means true and `#f` means false:

```
> (pair? (cons 1 2))
#t

> (list? (cons 1 2))
#f

> (pair? (cons 1 '()))
#t

> (list? (cons 1 '()))
#t
```

From this we can see that a list is always a pair, but the converse is not always true: a pair is not always a list.

Typically, `cons` is used to add an atomic value to the beginning of a list, like so:

```
> (cons 1 '(2 3))
'(1 2 3)
```

Racket provides special functions to access the components of a cons cell. The function `car` returns the item being pointed to by the car pointer, and correspondingly the `cdr` function returns the item being pointed to by the cdr pointer. In Racket the functions `first` and `rest` are similar to `car` and `cdr` but are not aliases for these functions, since they only work with lists. A few examples are given below.

```
> (car '(1 ("two" "three") 4 5))
1

> (first '(1 ("two" "three") 4 5))
1
```

```
> (cdr '(1 ("two" "three") 4 5))
'(("two" "three") 4 5)

> (rest '(1 ("two" "three") 4 5))
'(("two" "three") 4 5)
```

List elements can also be accessed with the functions `second`, `third`, . . . , `tenth`.

```
> (first '(1 2 3 4))
1

> (second '(1 2 3 4))
2

> (third '(1 2 3 4))
3
```

Finally, a value at any position can be extracted by using `list-ref`.

```
> (list-ref '(a b c) 0)
'a

> (list-ref '(a b c) 1)
'b
```

The `list-ref` function takes a list and the index of the value you want, with the list coming first. Notice that Racket uses *zero-based indexes*, meaning for any sequence of values, the first value has an index of 0, the second value has an index of 1, and so on.

A Few Useful List Functions

Let's quickly go through a number of useful list functions.

length

To get the length of a list, you can use the `length` function, like so:

```
> (length '(1 2 3 4 5))
5
```

reverse

If you need the elements in a list reversed, you can use the `reverse` function.

```
> (reverse '(1 2 3 4 5)) ; reverse elements of a list
'(5 4 3 2 1)
```

sort

The sort function will sort a list. You can pass in < to sort the list in ascending order:

```
> (sort '(1 3 6 5 7 9 2 4 8) <)
'(1 2 3 4 5 6 7 8 9)
```

Or, if you pass in >, it will sort the list in descending order:

```
> (sort '(1 3 6 5 7 9 2 4 8) >)
'(9 8 7 6 5 4 3 2 1)
```

append

To merge two lists together, you can use the append function:

```
> (append '(1 2 3) '(4 5 6))
'(1 2 3 4 5 6)
```

The append function can take more than two lists:

```
> (append '(1 2) '(3 4) '(5 6))
'(1 2 3 4 5 6)
```

range

The range function will create a list of numbers given some specifications. You can pass a start value and an end value, as well as a step to increment:

```
> (range 0 10 2)
'(0 2 4 6 8)
```

Or, if you just pass an end value, it will start at 0 with a step of 1:

```
> (range 10)
'(0 1 2 3 4 5 6 7 8 9)
```

make-list

Another way to make lists is using the make-list function:

```
> (make-list 10 'me)
'(me me me me me me me me me me)
```

As you can see, make-list takes a number and a value, and makes a list that contains that value repeated that number of times.

null?

To test whether a list is empty or not, you can use the null? function:

```
> (null? '()) ; test for empty list
#t
```

```
> (null? '(1 2 3))
#f
```

index-of

If you need to search a list for a value, you can use `index-of`. It'll return the index of the value if it appears:

```
> (index-of '(8 7 1 9 5 2) 9)
3
```

It'll return `#f` if it doesn't:

```
> (index-of '(8 7 1 9 5 2) 10)
#f
```

member

Another way to search lists is to use `member`, which tests whether a list contains an instance of a particular element. It returns the symbol `#f` if it does not, and returns the tail of the list starting with the first instance of the matching element if it does.

```
> (member 7 '(9 3 5 (6 2) 5 1 4))
#f

> (member 5 '(9 3 5 (6 2) 5 1 4))
'(5 (6 2) 5 1 4)

> (member 6 '(9 3 5 (6 2) 5 1 4))
#f
```

Notice that in the last instance, even though 6 is a member of a sublist of the searched list, the `member` function still returns false. However, the following does work.

```
> (member '(6 2) '(9 3 5 (6 2) 5 1 4))
'((6 2) 5 1 4)
```

You'll see later that in functional programming, you often need to determine whether an item is contained in a list. The `member` function not only finds the item (if it exists) but returns the actual value so that it can be used in further computations.

We'll have much more to say about lists in the remainder of this text.

Defines, Assigns, and Variables

Thus far, we've seen a few examples of a *function*, something that takes one or more input values and provides an output value (some form of data). The first element in a function-call expression is an identifier (the function name). The remaining elements in a function form are the arguments to the function. These elements are each evaluated and then fed to the function, which performs some operation on its arguments and returns a value.

More specifically, a *form* or *expression* may define a function, execute a function call, or simply return a structure (normally a list), and may or may not evaluate all its arguments. Notice that quote is a different type of form (distinct from a function form, which evaluates its arguments) since it *does not* first evaluate its arguments. In the next section you'll meet define, which is yet another type of form since it does not evaluate its first argument, but it does evaluate its second argument. We will meet many other types of forms as we progress through the text.

A *variable* is a placeholder for a value. In Racket, variables are specified by *identifiers* (specific sequences of characters) associated with one thing only. (We'll have more to say about what constitutes a valid identifier shortly.) To define a variable, you use the define form. For example:

```
> (define a 123)
> a
123
```

Here define is said to *bind* the value 123 to the identifier a. Virtually anything can be bound to a variable. Here we'll bind a list to the identifier b.

```
> (define b '(1 2 3))
> b
'(1 2 3)
```

It's possible to bind several variables in parallel:

```
> (define-values (x y z) (values 1 2 3))

> x
1

> y
2

> z
3
```

Racket makes a distinction between *defining* a variable and *assigning* a value to a variable. Assignments are made with a set! expression. Typically any form which changes, or *mutates*, a value will end with an exclamation point. Attempting to assign to an identifier that hasn't been previously defined will result in an ugly error message:

```
> (set! ice 9)
. . set!: assignment disallowed;
 cannot set variable before its definition
 variable: ice
```

But this is okay:

```
> (define ice 9)
> ice
9
> (set! ice 32)
32
```

One way to think of this is that define sets up a location to store a value, and set! simply places a new value in a previously defined location.

When we speak of a variable x that is defined in Racket code, it will be typeset as x. If we're simply speaking of the variable in the mathematical sense, it will be typeset in italics as x.

Symbols, Identifiers, and Keywords

Unlike most languages, Racket allows just about any string of characters to be used as an identifier. For example we can use 2x3 as an identifier:

```
> (define 2x3 7)
> 2x3
7
```

You could conceivably define a function literally called rags->riches that would convert rags to riches (let me know when you get that working). All this seems quite bizarre, but it lends Racket an expressive power not found in many other computer languages. There are of course some restrictions to this, but aside from a few special characters such as parentheses, brackets, and arithmetic operators (even these are usually okay if they aren't the first character), just about anything goes. In fact it's common to see identifiers containing dashes, as in solve-for-x.

A *symbol* is essentially just a quoted identifier:

```
> 'this-is-a-symbol
'this-is-a-symbol
```

They are sort of a second-rate string (more on strings below). They are typically used much like an enum in other programming languages where they're used to stand for a specific value.

A *keyword* is an identifier prefixed with #:. Keywords are mainly used to identify optional arguments in function calls. Here's an example of a function (~r) that uses a keyword to output π as a string with two decimal places of accuracy.

```
> (~r pi #:precision 2)
"3.14"
```

Here we define the optional precision argument to specify that the value of pi should be rounded to two decimal places.

Equality

Racket defines two different kinds of equality: things that look exactly alike and things that are the same thing. Here's the difference. Suppose we make the following two definitions.

```
> (define a '(1 2 3))
> (define b '(1 2 3))
```

Identifiers a and b look exactly alike, and if we ask Racket if they are the same with the equal? predicate, it will respond that they are the same. Note a *predicate* is a function that returns a Boolean value of true or false.

```
> (equal? a b)
#t
```

But if we ask whether they are the same thing by using the eq? predicate, we get a different answer.

```
> (eq? a b)
#f
```

So when does eq? return true? Here's an example.

```
> (define x '(1 2 3))
> (define y x)
> (eq? x y)
#t
```

In this case we have bound x to the list '(1 2 3). We then bind y to the same value *location* that x is bound to, effectively making x and y be bound to the same thing. The difference is subtle, but important. In most cases equal? is what you need, but there are scenarios where eq? is used to ensure that variables are bound to the same object and not just to things that *look* the same.

One other nuance of equality that must be discussed is numeric equality. In the discussion above, we were focused on structural equality. Numbers are a different animal. We'll have much more to say about numbers in the next chapter, but we need to clarify a few things about numbers that relate to equality. Examine the following sequence:

```
> (define a  123)
> (define b  123)
> (eq? a b)
#t
```

Above we bound a and b to identical lists '(1 2 3), and in that case eq? returned false. In this case we bound a and b to the identical number 123, and eq? returned true. Numbers (technically *fixnums*, that is, small integers that fit into a fixed amount of storage—typically 32 or 64 bits, depending on your computing platform) are unique in this sense. There is only one instance of every number, no matter how many different identifiers it is bound to. In other words, each number is stored in one and only one location. Furthermore, there's a special predicate (=) that can only be used with numbers:

```
> (= 123 123)
#t

> (= 123 456)
#f

(= '(1 2 3) '(1 2 3))
. . =: contract violation
  expected: number?
  given: '(1 2 3)
  argument position: 1st
  other arguments...:
```

In this section we only cover equality in general. We'll look at more specifics on numerical comparisons in the next chapter.

Strings and Things

In this section, we'll look at different ways of handling text values in Racket. We'll begin with the simplest kind of text value.

Characters

Individual text values, like single letters, are represented using a *character*, a special entity that corresponds to a *Unicode* value. For example, the letter A corresponds to the Unicode value 65. Unicode values are usually specified in hexadecimal, so the Unicode value for A is $65_{10} = 0041_{16}$. Character values either start with `#\` followed by a literal keyboard character or `#\u` followed by a Unicode value.

Here's a sampling of the multiple ways to write a character using character functions. Notice the use of the comment character (`;`), which allows comments (non-compiled text) to be added to Racket code.

```
> #\A
#\A

> #\u0041
#\A

> #\   ; this is a space character
#\space

> #\u0020  ; so is this
#\space
```

```
> (char->integer #\u0041)
65

> (integer->char 65)
#\A

> (char-alphabetic? #\a)
#t

> (char-alphabetic? #\1)
#f

> (char-numeric? #\1)
#t

> (char-numeric? #\a)
#f
```

Unicode supports a wide range of characters. Here are some examples:

```
> '(#\u2660 #\u2663 #\u2665 #\u2666)
'(#♠ #♣ #♡ #♢)

> '(#\u263A #\u2639 #\u263B)
'(\#☺ \#☹ \#☻)

> '(#\u25A1 #\u25CB #\u25C7)
'(\#□ \#○ \#◇)
```

Most Unicode characters should print fine, but this depends to some extent on the fonts available on your computer.

Strings

A *string* typically consists of a sequence of keyboard characters surrounded by double-quote characters.

```
> "This is a string."
"This is a string."
```

Unicode characters can be embedded in a string, but in this case, the leading # is left off.

```
> "Happy: \u263A."
"Happy: ☺."
```

You can also use `string-append` on two strings to create a new string.

```
> (string-append "Luke, " "I am " "your father!")
"Luke, I am your father!"
```

To access a character within a string, use `string-ref`:

```
> (string-ref "abcdef" 2)
#\c
```

The position of each character in a string is numbered starting from 0, so in this example using an index of 2 actually returns the third character.

The strings we have seen so far are immutable. To create a mutable string, use the `string` function. This allows changing characters in the string.

```
> (define wishy-washy (string #\I #\  #\a #\m #\  #\m #\u #\t #\a #\b #\l #\e)
    )
> wishy-washy
"I am mutable"

> (string-set! wishy-washy 5 #\a)
> (string-set! wishy-washy 6 #\ )

> wishy-washy
"I am a table"
```

Note that for mutable strings we have to define the string using individual characters.

Another way to create a mutable string is with `string-copy`:

```
> (define mstr (string-copy "I am also mutable"))
> (string-set! mstr 5 #\space)
> (string-set! mstr 6 #\space)
> mstr
"I am   so mutable"
```

You can also use `make-string` to do the same thing:

```
> (define exes (make-string 10 #\X))
> (string-set! exes 5 #\O)
> exes
"XXXXXOXXXX"
```

Depending on what's needed, any one of the above may be preferred. If you need to make an existing string mutable, `string-copy` is the obvious choice. If you only want a string of spaces, `make-string` is the clear winner.

Useful String Functions

There are of course a number of other useful string functions, a few of which we illustrate next.

string-length

The string-length function outputs the number of characters in a string (see wishy-washy earlier in "Strings" on page 14.

```
> (string-length wishy-washy)
12
```

substring

The substring function extracts a substring from a given string.

```
> (substring wishy-washy 7 12) ; characters 7-11
"table"
```

string-titlecase

Use string-titlecase to capitalize the first character of each word in a string.

```
> (string-titlecase wishy-washy)
"I Am A Table"
```

string-upcase

To output a string in all caps, use string-upcase:

```
> (string-upcase "big")
"BIG"
```

string-downcase

Conversely, for a lowercase string, use string-downcase:

```
> (string-downcase "SMALL")
"small"
```

string<=?

To perform an alphabetical comparison, use the string<=? function:

```
> (string<=? "big" "small")  ; alphabetical comparison
#t
```

string=?

The string=? function tests whether two strings are equal:

```
> (string=? "big" "small")
#f
```

string-replace

The string-replace function replaces part of a string with another string:

```
> (define darth-quote "Luke, I am your father!")
> (string-replace darth-quote "am" "am not")
"Luke, I am not your father!"
```

string-contains?

To test whether one string is contained within another, use string-contains?:

```
> (string-contains? darth-quote "Luke")
#t

> (string-contains? darth-quote "Darth")
#f
```

string-split

The string-split function can be used to split a string into tokens:

```
> (string-split darth-quote)
'("Luke," "I" "am" "your" "father!")

> (string-split darth-quote ",")
'("Luke" " I am your father!")
```

Notice that the first example above uses the default version that splits on spaces whereas the second version explictly uses a comma (,).

string-trim

The string-trim function gets rid of any leading and/or trailing spaces:

```
> (string-trim "  hello   ")
"hello"

> (string-trim "  hello   " #:right? #f)
"hello   "

> (string-trim "  hello   " #:left? #f)
"  hello"
```

Notice in the last two versions, #:left? or #:right? is used to suppress trimming the corresponding side. The final #f argument (the default) is used to specify that only one match is removed from each side; otherwise all initial or trailing matches are trimmed.

For more advanced string functionality, see "Regular Expressions" on page 279.

String Conversion and Formatting Functions

There are a number of functions that convert values to and from strings. They all have intuitive names and are illustrated below.

```
> (symbol->string 'FBI)
"FBI"

> (string->symbol "FBI")
'FBI

> (list->string '(#\x #\y #\z))
"xyz"

> (string->list "xyz")
'(#\x #\y #\z)

> (string->keyword "string->keyword")
'#:string->keyword

> (keyword->string '#:keyword)
"keyword"
```

For a complete list of these, go to *https://docs.racket-lang.org/reference/strings.html.*

A handy function to embed other values within a string is `format`.

```
> (format "let ~a = ~a" "x" 2)
"let x = 2"
```

Within the format statement, ~a acts as a placeholder. There should be one placeholder for each additional argument. Note that the number 2 is automatically converted to a string before it's embedded in the output string.

If you want to simply convert a number to a string, use the `number->string` function:

```
> (number->string pi)
"3.141592653589793"
```

Conversely:

```
> (string->number "3.141592653589793")
3.141592653589793
```

Trying to get Racket to translate the value of words into numbers, however, will not work:

```
> (string->number "five")
#f
```

For more control we can use the `~r` function, defined in the *racket/format* library, which has many options that can be used to convert a number to a string and control the precision and other output characteristics of the number. For example, to show π to four decimal places, we would use this:

```
> (~r pi #:precision 4)
"3.1416"
```

To show this right-justified, in a field 20 characters wide, and left padded with periods, we execute the following:

```
> (~r pi #:min-width 20 #:precision 4 #:pad-string ".")
".............3.1416"
```

Additional info on `~r` is available in Appendix A, which talks about number bases. There are a number of other useful tilde-prefixed string conversion functions available, such as `~a`, `~v`, and `~s`. We won't go into detail here, but you can consult the Racket Documentation for details: *https://docs.racket-lang.org/reference/strings.html*.

Vectors

Vectors bear a superficial resemblance to lists, but they are quite different. In contrast to the internal tree structure of lists, *vectors* are a sequential array of cells (much like arrays in imperative languages) that directly contain values, as illustrated in Figure 1-3.

1	3	"d"	'a	2
0	1	2	3	4

Figure 1-3: Vector structure

Vectors can be entered using the `vector` function.

```
> (vector 1 3 "d" 'a 2)
'#(1 3 "d" a 2)
```

Alternatively, vectors can be entered using `#` as follows (note that an unquoted `#` implies a quote):

```
> #(1 3 "d" a 2)
'#(1 3 "d" a 2)
```

It's important to note that these methods are *not* equivalent. Here's one reason why:

```
> (vector 1 2 pi)
'#(1 2 3.141592653589793)

> #(1 2 pi)
'#(1 2 pi)
```

In the first example, just as for `list`, `vector` first evaluates its arguments before forming the vector. In the last example, like `quote`, `#` does not evaluate its arguments. More importantly, `#` is an alias for `vector-immutable`, which leads to our next topic.

Accessing Vector Elements

The function `vector-ref` is an indexing operator that returns an element of a vector. This function takes a vector as its first argument and an index as its second:

```
> (define v (vector 'alpha 'beta 'gamma))
> (vector-ref v 1)
'beta

> (vector-ref v 0)
'alpha
```

To assign a value to a vector cell, `vector-set!` is used. The `vector-set!` expression takes three arguments: a vector, an index, and a value to be assigned to the indexed position in the vector.

```
> (vector-set! v 2 'foo)
> v
'#(alpha beta foo)
```

Let's try this a bit differently:

```
> (define u #(alpha beta gamma))
> (vector-set! u 2 'foo)
. . vector-set!: contract violation
  expected: (and/c vector? (not/c immutable?))
  given: '#('alpha 'beta 'gamma)
  argument position: 1st
  other arguments...:
```

Remember that `#` is an alias for `vector-immutable`. What this means is that vectors created with `#` (or `vector-immutable`) are (drum roll . . .) *immutable*: they cannot be changed or assigned new values. On the other hand, vectors created with `vector` are *mutable*, meaning that their cells can be modified.

One advantage of vectors over lists is that elements of vectors can be accessed much faster than elements of lists. This is because to access the 100th element of a list, each cell of the list must be accessed sequentially to get to the 100th element. Conversely, with vectors, the 100th element can be accessed directly, without working through earlier cells. On the other hand, lists are quite flexible and can easily be extended as well as being used to represent other data structures like trees. They are the bread and butter of Racket (and all Lisp-based languages), so much of the functionality of the language depends on the list structure. Predictably, functions are provided to easily convert from one to the other.

Useful Vector Functions

vector-length

The `vector-length` function returns the number of elements in a vector:

```
> (vector-length #(one ringy dingy))
3
```

vector-sort

The `vector-sort` function sorts the elements of a vector:

```
> (vector-sort #(9 1 3 8 2 5 4 0 7 6 ) <)
'#(0 1 2 3 4 5 6 7 8 9)

> (vector-sort #(9 1 3 8 2 5 4 0 7 6 ) >)
'#(9 8 7 6 5 4 3 2 1 0)
```

To whet your appetite for what's to come later, `vector-sort` is a typical example of functional programming. The last argument actually evaluates a function that determines the direction of the sort.

vector->list

The `vector->list` function takes a vector and returns a list:

```
>  (vector->list #(one little piggy))
'(one little piggy)
```

list->vector

Conversely `list->vector` takes a list and returns a vector:

```
> (list->vector '(two little piggies))
'#(two little piggies)
```

make-vector

To create a mutable vector, use the `make-vector` form:

```
> (make-vector 10 'piggies) ; create a mutable vector
'#(piggies piggies piggies piggies piggies piggies piggies piggies piggies
    piggies)
```

vector-append

To concatenate two vectors together, use `vector-append`:

```
> (vector-append #(ten little) #(soldier boys))
'#(ten little soldier boys)
```

vector-member

The `vector-member` function returns the index to where an item is located in a vector:

```
> (vector-member 'waldo (vector 'where 'is 'waldo '?) )
2
```

There are of course many other useful vector functions, and we will explore some of them in the chapters to come.

Using structs

To introduce the next Racket feature, let's build an example program. Instead of keeping your checkbook transactions in a paper bankbook, you could create an electronic version using Racket. Typically such transactions have the following components:

- Transaction date
- Payee
- Check number
- Amount

One way to keep track of these disparate pieces of information is in a Racket structure called a `struct`. A Racket `struct` is conceptually similar to a `struct` in languages such as C or C++. It's a composite data structure that has a set of predefined fields. Before you can use a `struct`, you have to tell Racket what it looks like. For our bank transaction example, such a definition might look like this:

```
> (struct transaction (date payee check-number amount))
```

Each of the components of a structure (`date`, `payee`, etc.) is called a *field*. Once we've defined our `transaction struct`, we can create one like this:

```
> (define trans (transaction 20170907 "John Doe" 1012 100.10))
```

Racket automatically creates an *accessor method* for each of the fields in the structure. An accessor method returns the value of the field. They always begin with the name of the structure (in this case `transaction`), a hyphen, and then the name of the field.

```
> (transaction-date trans)
20170907

> (transaction-payee trans)
"John Doe"

> (transaction-check-number trans)
1012
```

```
> (transaction-amount trans)
100.1
```

Suppose, however, that you made a mistake and determined that the check to John Doe should have been for $100.12 instead of $100.10 and try to correct it via `set-transaction-amount!`. Note the exclamation point: this is a signal that `set-transaction-amount!` is a *mutator*, that is, a method that modifies a field value). These mutators are generated when the struct is defined and typically start with set and end with `!`.

```
> (set-transaction-amount! trans 100.12)
. . set-transaction-amount!: undefined;
  cannot reference an identifier before its definition
```

Oops . . . Fields in a structure are immutable by default and hence do not export *mutators*. The way around this is to include the `#:mutable` keyword in the structure definition for any field that may need to be modified.

```
> (struct transaction
    (date payee check-number [amount #:mutable]))
> (define trans (transaction 20170907 "John Doe" 1012 100.10))
> (set-transaction-amount! trans 100.12)
> (transaction-amount trans)
100.12
{
```

If all the fields should be mutable, adding the `#:mutable` keyword after the field list will do the trick.

```
> (struct transaction
    (date payee check-number amount) #:mutable)
> (define trans (transaction 20170907 "John Doe" 1012 100.10))
> (set-transaction-check-number! trans 1013)
> (transaction-check-number trans)
1013
```

While the accessor methods are sufficient for getting the value of a single field, they are a bit cumbersome for seeing all the values at once. Just entering the structure name does not yield much information.

```
> trans
#<transaction>
```

To make your structure more transparent, include the `#:transparent` option in the `struct` definition.

```
> (struct transaction
    (date payee check-number amount) #:mutable #:transparent)
> (define trans (transaction 20170907 "John Doe" 1012 100.10))
> trans
(transaction 20170907 "John Doe" 1012 100.1)
```

There are additional useful options that can be applied when defining structures, but one that is of particular interest is `#:guard`. `#:guard` provides a mechanism to validate the fields when a structure is constructed. For instance, to ensure that negative check numbers are not used, we could do the following.

```
> (struct transaction
    (date payee check-number amount)
    #:mutable #:transparent
    #:guard (λ (date payee num amt name)
        (unless (> num 0)
        (error "Not a valid check number"))
        (values date payee num amt)))

> (transaction 20170907 "John Doe" -1012 100.10)
Not a valid check number

> (transaction 20170907 "John Doe" 1012 100.10)
(transaction 20170907 "John Doe" 1012 100.1)
```

Don't panic. We haven't covered that funny-looking symbol (λ, or *lambda*) yet, but you should be able to get the gist of what's going on. The `#:guard` expression is a function that takes one parameter for each field and one additional parameter that contains the structure name. In this case we're only testing whether the check number is greater than zero. The `#:guard` expression must return the same number of values as the number of fields in the `struct`.

In the previous example we simply returned the same values that were entered, but suppose we had a variable that contained the last check number called `last-check`. In this case, we could enter a 0 for the check number and use the `#:guard` expression to plug in the next available number as shown here.

```
> (define last-check 1000)

> (struct transaction
    (date payee check-number amount)
    #:mutable #:transparent
    #:guard (λ (date payee num amt name)
              (cond
                [(< num 0)
                   (error "Not a valid check number")]
                [(= num 0)
                   (let ([next-num (add1 last-check)])
                     (set! last-check next-num)
                     (values date payee next-num amt))]
                [else
                   (set! last-check num)
                   (values date payee num amt)])))
```

```
> (transaction 20170907 "John Doe" 0 100.10)
(transaction 20170907 "John Doe" 1001 100.1)

> (transaction 20170907 "Jane Smith" 1013 65.25)
(transaction 20170907 "Jane Smith" 1013 65.25)

> (transaction 20170907 "Acme Hardware" 0 39.99)
(transaction 20170907 "Acme Hardware" 1014 39.99)
```

As you can see, non-zero check numbers are stored as the last check number, but if a zero is entered for the check number, the `struct` value gets generated with the next available number, which becomes the current value for `last-check`. The `cond` statement will be explained in more detail a bit later in the book, but its use here should be fairly clear: it's a way to check multiple cases.

Controlling Output

In the interactions pane, DrRacket immediately displays the output resulting from evaluating any expression. It's often desirable to have some control over how the output is presented. This is especially important when the output is being generated by some function or method. Racket provides a number of mechanisms for generating formatted output. The main forms are `write`, `print`, and `display`. Each of these works in a slightly different way. The best way to illustrate this is with examples.

write

The `write` expression outputs in such a way that the output value forms a valid value that can be used in the input:

```
> (write "show me the money")
"show me the money"

> (write '(show me the money))
(show me the money)

> (write #\A)
#\A

> (write 1.23)
1.23

> (write 1/2)
1/2

> (write #(a b c))
#(a b c)
```

display

The `display` expression is similar to `write`, but strings and character data types are written as raw strings and characters without any adornments such as quotation or tick marks:

```
> (display "show me the money")
show me the money

> (display '(show me the money))
(show me the money)

> (display #\A)
A

> (display 1.23)
1.23

> (display 1/2)
1/2

> (display #(a b c))
#(a b c)
```

print

The `print` expression is also similar to `write`, but adds a bit more formatting to the output. The intent of `print` is to show an expression that would evaluate to the same value as the printed one:

```
> (print "show me the money")
"show me the money"

> (print '(show me the money))
'(show me the money)

> (print #\A)
#\A

> (print 1.23)
1.23

> (print 1/2)
1
-
2

> (print #(a b c))
'#(a b c)
```

Notice how the rational value `1/2` is printed (more on rationals in the next chapter).

Each of these comes in a form that ends with `ln`. The only difference is that the ones that end with `ln` automatically print a new line at the end. Here are a couple of examples to highlight the difference.

```
> (print "show me ") (print "the money")
"show me ""the money"

> (display "show me ") (display "the money")
show me the money

> (println "show me ") (println "the money")
"show me "
"the money"

> (displayln "show me ") (displayln "the money")
show me
the money
```

One very useful form is `printf`. The `printf` expression works much like the `format` function: it takes a format string as its first argument and any number of other values as its other argument. The format string uses `~a` as a placeholder. There must be one placeholder for each of the arguments after the format string. The format string is printed exactly as entered, with the exception that for each placeholder the corresponding argument is substituted. Here's `printf` in action.

```
> (printf "~a + ~a = ~a" 1 2 (+ 1 2))
1 + 2 = 3

> (printf "~a, can you hear ~a?" "Watson" "me")
Watson, can you hear me?

> (printf "~a, can you hear ~a?" "Jeeves" "the bell")
Jeeves, can you hear the bell?
```

There are additional format specifiers (see the Racket Documentation for details), but we'll mostly be using `print` since it gives a better visual indication of the data type of the value being output.

Summary

In this chapter, we laid the groundwork for what's to come. Most of the core data types have been introduced along with what are hopefully some helpful examples. By now you should be comfortable with basic Racket syntax and have a pretty good understanding of the structure of lists and how to manipulate them. The next chapter will take a detailed look at the various numeric data types provided by Racket.

2

ARITHMETIC AND OTHER NUMERICAL PARAPHERNALIA

In this chapter, we'll take a look at the rich set of numerical data types that Racket provides. We'll discover the expected integer and floating-point values, but we'll also learn that Racket supports rational (or fractional) values along with complex numbers (don't worry if you don't know what complex numbers are; they are not heavily used in this text, but we take a brief look for those that may be interested).

Booleans

Booleans are true and false values, and while they aren't strictly numbers, they behave a bit like numbers in that they can be combined by various operators to produce other Boolean values. The discipline governing these operations is known as *Boolean algebra*. In Racket, Booleans are represented by the values `#t` and `#f`, true and false respectively. It's also possible to use `#true` (or `true`) and `#false` (or `false`) as aliases for `#t` and `#f` respectively.

Before we introduce specific Boolean operators, one important observation about Racket Boolean operators in general is that they typically treat any value that's not literally #f as true. You'll see some examples of this behavior below.

The first operator we'll look at is not, which simply converts #t to #f and vice versa.

```
> (not #t)
#f

> (not #f)
#t

> (not 5)
#f
```

Notice that 5 was converted to #f, meaning that it was originally treated as #t.

The next Boolean operator we'll look at is and, which returns true if all its arguments are true. Let's look at some examples:

```
> (and #t #t)
#t

> (and #t #f)
#f

> (and 'apples #t)
#t

> (and (equal? 5 5) #f)
#f

> (and (equal? 5 5) #t)
#t

> (and (equal? 5 5) #t 23)
23
```

You may be a bit puzzled by the last example (and rightfully so). Remember that Racket considers all non-false values as true, so 23 is in fact a valid return value. More important though is how and evaluates its arguments. What happens in reality is that and sequentially evaluates its arguments until it hits a #f value. If no #f value is encountered, it returns the value of its last argument, 23 in the example above. While this behavior seems a bit odd, it is consistent with how the or operator works, where, as we'll see shortly, it can be quite useful in certain circumstances.

The last Boolean operator we'll look at is the `or` operator, which will return true if any of its arguments are true and `#f` otherwise. Here are some examples:

```
> (or #f #f)
#f

> (or #f #t)
#t

> (or #f 45 (= 1 3))
45
```

Much like `and`, `or` sequentially evaluates its arguments. But in `or`'s case, the first *true* value is returned. In the example above, 45 is treated as true, so that's the value returned. This behavior can be quite useful when one wants the first value that's not `#f`.

Other less frequently used Boolean operators are `nand`, `nor`, and `xor`. Consult the Racket Documentation for details on these operators.

The Numerical Tower

In mathematics there's a hierarchy of number types. *Integers* are a subset of rational (or fractional) numbers. *Rational numbers* are a subset of real numbers (or floating-point values as they are approximated by computers). And *real numbers* are a subset of complex numbers. This hierarchy is known as the *numerical tower* in Racket.

Integers

In mathematics the set of integers is represented by the symbol $\mathbb{Z}$. Racket integers consist of a sequence of digits from 0 to 9, optionally preceded by a plus or minus sign. Integers in Racket are said to be *exact*. What this means is that applying arithmetical operations to exact numbers will always produce an exact numerical result (in this case a number that's still an integer). In many computer languages, once an operation produces a number of a certain size, the result will either be incorrect or it will be converted to an approximate value represented by a floating- point number. With Racket, numbers can get bigger and bigger until your computer literally runs out of memory and explodes. Here are some examples.

```
> (+ 1 1)
2

> (define int 1234567890987654321)
> (* int int int int)
2323057235416375647706123102514602108949250692331618011140356079618623681

> (- int)
```

```
-1234567890987654321

> (- 5 -7)
12

> (/ 4 8)
1/2

> (/ 5)
1/5
```

Note that in the last examples, division doesn't result in a floating-point number but rather returns an *exact* value: a rational number (discussed in the next section).

It's possible to enter integers in number bases other than 10. Racket understands *binary numbers* (integers prefixed by `#b`), *octal* numbers (integers prefixed by `#o`), and *hexadecimal* numbers (integers prefixed by `#x`):

```
> #b1011
11

> #b-10101
-21

> #o666
438

> #xadded
712173
```

Non-decimal bases have somewhat specialized use cases, but one example is that HTML web pages typically express color values as hexadecimal numbers. Also, binary numbers are how computers store all values internally, so they can be useful for individuals studying basic computer science. Octal and hexadecimal values have a further advantage: binary numbers can easily be converted to octal since three binary digits equates to a single octal value and four binary digits equates to a single hexadecimal digit.

Rationals

Next up on the mathematical food chain are the rational numbers (or fractions), expressed by the mathematical symbol $\mathbb{Q}$. Fractions in Racket consist of two positive integer values separated by a forward slash (no spaces allowed), optionally preceded by a plus or minus sign. Rational numbers are also an exact numeric type, and all operations permitted for integers are also valid for rational numbers.

```
> -2/4
-1/2

> 4/6
2/3

> (+ 1/2 4/8)
1

> (- 1/2 2/4 4/8 8/16)
-1

> (* 1/2 2/3)
1/3

> (/ 2 2/3)
3
```

The numerator and denominator of a rational number can be obtained with the `numerator` and `denominator` functions.

```
> (numerator 2/3)
2

> (denominator 2/3)
3
```

Reals

A *real* number is a mathematical concept (specified by the symbol $\mathbb{R}$) that, in reality, does not exist in the world of computers. Real numbers such as π have an infinite decimal expansion that can only be approximated in a computer. Thus, we reach our first class of *inexact* numbers: floating-point numbers. Floating-point numbers in Racket are entered in the same way as they are in most programming languages and calculators. Here are some (unfortunately boring) examples:

```
> -3.14159
-3.14159

> 3.14e159
3.14e+159

> pi
3.141592653589793

> 2.718281828459045
```

```
2.718281828459045

> -20e-2
-0.2
```

It's important to keep in mind that there are some subtle distinctions in the mathematical concept of certain number types and what they mean in a computing environment. For example a number entered as 1/10 is, as mentioned above, treated as an exact rational number since it can be represented as such in a computer (internally it's stored as two binary integer values), but the value 0.1 is treated as an inexact floating-point value, an approximation of the real number value, since it cannot be represented internally a single binary value (at least not without using an infinite number of binary digits).

Complex Numbers

When we use the term *complex* number it does not mean we are speaking of a *complicated* number, but rather a special type of number. If you're not already familiar with this concept, there's no harm in moving on to the next section, since complex numbers aren't used in the remainder of the book (although I would encourage you to read up on this fascinating subject). This section is included as a reference for the brave souls who may make use of this information in their own projects.

Complex numbers are entered almost exactly as they appear in any mathematical text, but there are some points to note. First, if the real component is omitted, the imaginary part must be preceded by a plus or minus sign. Second, there can be no spaces in the string used to define the number. And third, complex numbers must end in `i`. Examples:

```
> +1i ; our friend, the imaginary number
0+1i

> 1i ; this will give an error
. . 1i: undefined;
  cannot reference an identifier before its definition

> +i ; it is even possible to leave off the 1
0+1i

> -1-234i
-1-234i

> -1.23+4.56i
-1.23+4.56i

> 1e10-2e10i
10000000000.0-20000000000.0i
```

Note that complex numbers can be exact or inexact. We can test exactness using the exact? operator:

```
> (exact? 1/2+8/3i)
#t

> (exact? 0.5+8/3i)
#f
```

To get at the components of a complex number, use `real-part` and `imag-part`:

```
> (real-part 1+2i)
1

> (imag-part 1+2i)
2
```

This concludes our look at the numerical tower and basic arithmetical operations on the various number types. In the next few sections we'll look at comparison operators, what happens when different number types are added together (for example adding an integer to a floating-point number), and some useful mathematical functions.

Numeric Comparison

Racket supports the usual complement of numeric comparison operators. We can test if numbers are equal:

```
> (= 1 1.0)
#t

> (= 1 2)
#f

> (= 0.5 1/2)
#t
```

and compare their sizes:

```
> (< 1 2)
#t

> (<= 1 2)
#t

> (>= 2 1.9)
#t
```

You can also use these operators on multiple arguments, and Racket will ensure that the elements pair-wise satisfy the comparison operator. In the example below, this means that 1 < 2, 2 < 3, and 3 < 4.

```
> (< 1 2 3 4)
#t

> (< 1 2 4 3)
#f
```

But there's no *not equals* operator, so to test if two numbers are not equal to each other, you would have to do something like the following:

```
> (not (= 1 2))
#t
```

Combining Data Types

As you saw above, you can compare numbers of different types. But notice that we only performed arithmetic on exact numbers with exact numbers and vice versa. Here we'll discuss the implications of mixing exact and inexact numbers. Mixing exact and inexact numbers won't result in mass chaos (think *Ghostbusters* stream-crossing), but there are some fine points you should be aware of.

First and foremost, when it comes to arithmetic operators (addition, subtraction, and so on), the rules are fairly simple:

Mixing exact with exact will give an exact result.

Mixing inexact with inexact will give an inexact result.

Mixing exact with inexact (or vice versa) will give an inexact result.

No surprises here, but there are some nuanced exceptions to these rules, such as multiplying anything by zero gives exactly zero.

Trigonometric functions will generally always return an inexact result (but again, there are some reasonable exceptions; for example `exp 0` gives an exact 1). You'll see some of these functions later in the chapter. The square function, `sqr`, will return an exact result if given an exact number. Its square root counterpart, `sqrt`, will return an exact result if it's given an exact number *and* the result is an exact number; otherwise, it will return an inexact number:

```
> (sqrt 25)
5

> (sqrt 24)
4.898979485566356

> (sqr 1/4)
1/16
```

```
> (sqr 0.25)
0.0625

> (sqrt 1/4)
1/2

> (sqrt -1)
0+1i
```

There are a couple of functions available to test exactness. Earlier you saw the function exact?, which returns #t if its argument is an exact number; otherwise it returns #f. Its counterpart is inexact?. It's also possible to force an exact number to be inexact and vice versa using two built-in functions:

```
> (exact->inexact 1/3)
0.3333333333333333

> (inexact->exact pi)
3 39854788871587/281474976710656
>
```

There's a predicate to test for each of the numeric data types we have mentioned in this section, but they may not work exactly as you expect.

```
> (integer? 70)
#t

> (real? 70.0)
#t

> (complex? 70)
#t

> (integer? 70.0)
#t

> (integer? 1.5)
#f

> (rational? 1.5)
#t

> (rational? 1+5i)
#f

> (real? 2)
#t
```

```
> (complex? 1+2i)
#t
```

These predicates return a result that honors the mathematical meaning of the predicate. You may have expected (complex? 70) to return #f, but integers are complex numbers, just with a zero real component. Likewise, you may have expected (integer? 70.0) to return #f since it's a floating-point number, but since the fractional part is 0, the number (while also real) is in fact an integer (but not an exact number). The number 1.5 is equivalent to 3/2, so Racket considers this to be a rational number (but again, inexact). The number type predicates (integer?, rational?, real?, and complex?) are aligned with the mathematical hierarchy (or numerical tower) as mentioned at the beginning of the section.

Built-in Functions

Aside from the normal arithmetical operators illustrated above, Racket provides the usual complement of mathematical functions that are standard fare in any programming language. A generous litany of examples follows.

```
> (abs -5)
5

> (ceiling 1.5)
2.0

> (ceiling 3/2)
2

> (floor 1.5)
1.0

> (tan (/ pi 4))
0.9999999999999999

> (atan 1/2)
0.4636476090008061

> (cos (* 2 pi))
1.0

> (sqrt 81)
9

> (sqr 4)
16

> (log 100) ; natural logarithm
```

```
4.605170185988092

> (log 100 10) ; base 10 logarithm
2.0

> (exp 1) ; e^1
2.718281828459045

> (expt 2 8) ; 2^8
256
```

Note that when possible, a function that has an exact argument will return an exact result.

There are of course many other functions available. Consult the Racket Documentation for details.

Infix Notation

As we've seen, in Racket, mathematical operators are given before the operands: `(+ 1 2)`. Typical mathematical notation has the operator between the operands: 1 + 2. This is called *infix notation*. Racket natively allows a form of infix notation by using a period operator. Here are some examples.

```
> (1 . >= . 2)
#f

> (1 . < . 2)
#t

> (1 . + . 2)
3

> (2 . / . 4)
1/2

> (2 . * . 3)
6
```

This can be useful when we want to make explicit the relationship between certain operators, but it's unwieldy for complex expressions.

For complex mathematical expressions, Racket provides the `infix` package. This package can be imported with the following code:

```
#lang at-exp racket
(require infix)
```

The `#lang` keyword allows us to define language extensions (in this case the `at-exp` allows us to use @-expressions, which we will see shortly). The `require infix` expression states that we want to use the *infix* library.

Unfortunately, the infix package is not installed by default and must be installed from the Racket package manager (the package manager can be accessed through the DrRacket File menu) or the raco command line tool (if the
executable for raco is not in your execution path, it can be launched directly from the Racket install folder). To install using raco, execute the following on the command line:

```
> raco pkg install infix
```

Also note that we're using the language extension at-exp, which, while not entirely necessary, provides a nicer syntax to enter infix expressions. For example without at-exp, to compute $1 + 2 * 3$, we would enter the following:

```
> ($ "1+2*3")
7
```

With the at-exp extension, we could enter this:

```
> @${1+2*3}
7
```

While this only saves a couple of keystrokes, it removes the annoying string delimiters and just looks a bit more natural.

Function calls are handled in a familiar way by using square brackets. For example

```
> @${1 + 2*sin[pi/2]}
3.0
```

There is even a special form for lists:

```
> @${{1, 2, 1+2}}
'(1 2 3)
```

And there's one for variable assignments (which use :=, equivalent to set!, so the variable must be bound first):

```
> (define a 5)
> @${a^2}
25
> @${a := 6}
> @${2*a + 7}
19
```

To further illustrate the capabilities of the infix package, below is a complete program containing a function called quad, which returns a list containing the roots of the quadratic equation

$$ax^2 + bx + c = 0$$

As you'll recall from your algebra class (you *do* remember, don't you), these roots are given by

$$x = \frac{-b \pm \sqrt{b^2 - 4ac}}{2a}$$

```
#lang at-exp racket
(require infix)

(define (quad a b c)
  (let ([d 0])
    @${d := sqrt[b^2 - 4 * a * c];
         {(-b + d)/(2*a), (-b - d)/(2*a)}}))
```

After compiling this, we can solve $2x^2 - 8x + 6 = 0$ for x, by entering

```
> @${quad[2, -8, 6]}
'(3 1)
```

or equivalently . . .

```
> (quad 2 -8 6)
'(3 1)
```

Summary

With these first two chapters under your belt, you should be thoroughly familiar with Racket's basic data types. You should also be comfortable performing mathematical operations in Racket's rich numerical environment. This should prepare you for the somewhat more interesting topics to follow where we will explore number theory, data analysis, logic programming, and more. But, next up is functional programming, where we get down to the nitty-gritty of actually creating programs.

3

FUNCTION FUNDAMENTALS

In the last chapter, we introduced you to Racket's basic numerical operations. In this chapter, we'll explore the core ideas that form the subject of functional programming.

What Is a Function?

A *function* can be thought of as a box with the following characteristics: if you push an object in one side, an object (possibly the same, or not) comes out the other side; and for any given input item, the same output item comes out. This last characteristic means that if you put a triangle in one side and a star comes out the other, the next time you put a triangle in, you will also get a star out (see Figure 3-1). Unfortunately, Racket doesn't have any built-in functions that take geometric shapes as input, so we'll need to settle for more-mundane objects like numbers or strings.

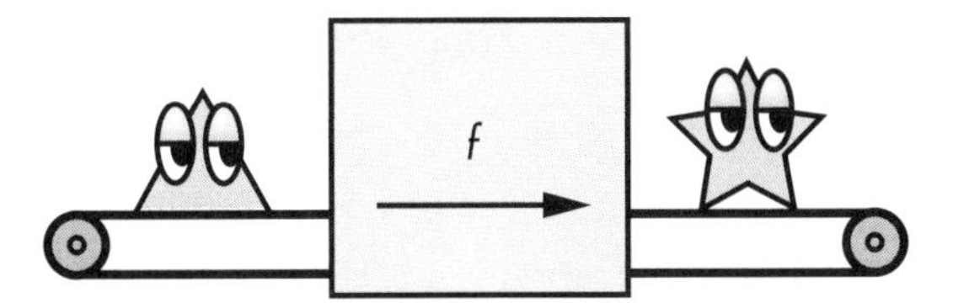

Figure 3-1: How a function works

Lambda Functions

In its most basic form, a function in Racket is something produced by a *lambda expression*, designated by the Greek letter λ. This comes from a mathematical discipline called lambda calculus, an arcane world we won't explore here. Instead, we'll focus on practical applications of lambda expressions. Lambda functions are intended for short simple functions that are immediately applied, and hence, don't need a name (they're anonymous). For example, Racket has a built-in function called `add1` that simply adds 1 to its argument. A Racket lambda expression that does the same thing looks like this:

```
(lambda (x) (+ 1 x))
```

Racket lets you abbreviate `lambda` with the Greek symbol λ, and we'll frequently designate it this way. You can enter λ in DrRacket by selecting it from the Insert menu or using the keyboard shortcut CTRL-\. We could rewrite the code above to look like this:

```
(λ (x) (+ 1 x))
```

To see a lambda expression in action, enter the following in the interactions pane:

```
> ((λ (x y) (+ (* 2 x) y)) 4 5)
13
```

Notice that instead of a function name as the first element of the list, we have the actual function. Here 4 and 5 get passed to the lambda function for evaluation.

An equivalent way of performing the above computation is with a `let` form.

```
> (let ([x 4]
        [y 5])
    (+ (* 2 x) y))
13
```

This form makes the assignment to variables `x` and `y` more obvious.

We can use lambda expressions in a more conventional way by assigning them to an identifier (a named function).

```
> (define foo (λ (x y) (+ (* 2 x) y)))
> (foo 4 5)
13
```

Racket also allows you to define functions using this shortcut:

```
> (define (foo x y) (+ (* 2 x) y))
> (foo 4 5)
13
```

These two forms of function definition are entirely equivalent.

Higher-Order Functions

Racket is a functional programming language. *Functional programming* is a programming paradigm that emphasizes a declarative style of programming without side effects. A *side effect* is something that changes the state of the programming environment, like assigning a value to a global variable.

Lambda values are especially powerful because they can be passed as values to other functions. Functions that take other functions as values (or return a function as a value) are known as *higher-order functions*. In this section, we'll explore some of the most commonly used higher-order functions.

The map Function

One of the most straightforward higher-order functions is the `map` function, which takes a function as its first argument and a list as its second argument, and then applies the function to each element of the list. Here's an example of the `map` function:

```
> (map (λ (x) (+ 1 x)) '(1 2 3))
'(2 3 4)
```

You can also pass a named function into `map`:

```
> (define my-add1 (λ (x) (+ 1 x)))
> (map my-add1 '(1 2 3)) ; this works too
'(2 3 4)
```

In the first example above, we take our increment function and pass it into `map` as a value. The `map` function then applies it to each element in the list `'(1 2 3)`.

It turns out that `map` is quite versatile. It can take as many lists as the function will accept as arguments. The effect is sort of like a zipper, where the list arguments are fed to the function in parallel, and the resulting values is a single list, formed by applying the function to the elements from each list. The example below shows `map` being used to add the corresponding elements of two equally sized lists together:

```
> (map + '(1 2 3) '(2 3 4))
'(3 5 7)
```

As you can see, the two lists were combined by adding the corresponding elements together.

The apply Function

The `map` function lets you apply a function to each item in a list individually. But sometimes, we want to apply all the elements of a list as arguments in a single function call. For example, Racket arithmetical operators can take multiple numeric arguments:

```
> (+ 1 2 3 4)
10
```

But if we try to pass in a list as an argument, we'll get an error:

```
> (+ '(1 2 3 4))
. . +: contract violation
  expected: number?
  given: '(1 2 3 4)
```

The + operator is only expecting numeric arguments. But not to worry. There's a simple solution: the apply function:

```
> (apply + '(1 2 3 4))
10
```

The apply function takes a function and a list as its arguments. It then *applies* the function to values in the list as if they were arguments to the function.

The foldr and foldl Functions

Yet another way to add the elements of a list together is with the foldr function. The foldr function takes a function, an initial argument, and a list:

```
> (foldr + 0 '(1 2 3 4))
10
```

Even though foldr produced the same result as apply here, behind the scenes it worked very differently. This is how foldr added the list together: $1 + (2 + (3 + (4 + 0)))$. The function "folds" the list together by performing its operation in a right-associative fashion (hence the r in foldr).

Closely associated with foldr is foldl. The action of foldl is slightly different from what you might expect. Observe the following:

```
> (foldl cons '() '(1 2 3 4))
'(4 3 2 1)

> (foldr cons '() '(1 2 3 4))
'(1 2 3 4)
```

One might have expected foldl to produce '(1 2 3 4), but actually foldl performs the computation (cons 4 (cons 3 (cons 2 (cons 1 '())))). The list arguments are processed from left to right, but the two arguments fed to cons are reversed—for example, we have (cons 1 '()) and not (cons '() 1).

The compose Function

Functions can be combined together, or *composed*, by passing the output of one function to the input of another. In math, if we have $f(x)$ and $g(x)$, they can be composed to make $h(x) = f(g(x))$ (in mathematics text this is sometimes designated with a special composition operator as $h(x) = (f \circ g)(x)$. We can do this in Racket using the compose function, which takes two or more functions and returns a new composed function. This new function works a bit like a pipeline. For example, if we want to increment a number by 1 and square the result (that is, for any n compute $(n + 1)^2$), we could use following function:

```
(define (n+1_squared n) (sqr (add1 n)))
```

But compose allows this to be expressed a bit more succinctly:

```
> (define n+1_squared (compose sqr add1))
> (n+1_squared 4)
25
```

Even simpler . . .

```
> ((compose sqr add1) 4)
25
```

Please note that add1 is performed first and then sqr. Functions are composed from right to left—that is, the rightmost function is applied first.

The filter Function

Our final example is filter. This function takes a predicate (a function that returns a Boolean value) and a list. The returned value is a list such that only elements of the original list that satisfy the predicate are included. Here's how we'd use filter to return the even elements of a list:

```
> (filter even? '(1 2 3 4 5 6))
'(2 4 6)
```

The filter function allows you to filter out items in the original list that won't be needed.

As you've seen throughout this section, our description of a function as a box is apt since it is in reality a value that can be passed to other functions just like a number, a string, or a list.

Lexical Scoping

Racket is a lexically scoped language. The Racket Documentation provides the following definition for *lexical scoping*:

> Racket is a lexically scoped language, which means that whenever an identifier is used as an expression, something in the textual environment of the expression determines the identifier's binding.

What's important about this definition is the term *textual environment*. A textual environment is one of two things: the *global environment*, or forms where identifiers are bound. As we've already seen, identifiers are bound in the global environment (sometimes referred to as the top level) with define. For example

```
> (define ten 10)
> ten
10
```

The values of identifiers bound in the global environment are available everywhere. For this reason, they should be used sparingly. Global definitions should normally be reserved for function definitions and constant values. This, however, is not an edict, as there are other legitimate uses for global variables.

Identifiers bound within a form will *normally* not be defined outside of the form environment (but see "Time for Some Closure" on page 58 for an intriguing exception to this rule).

Let's look at a few examples.

Previously we explored the lambda expression `((λ (x y) (+ (* 2 x) y )) 4 5)`. Within this expression, the identifiers `x` and `y` are bound to 4 and 5. Once the lambda expression has returned a value, the identifiers are no longer defined.

Here again is the equivalent `let` expression.

```
(let ([x 4]
      [y 5])
  (+ (* 2 x) y))
```

You might imagine that the following would work as well:

```
(let ([x 4]
      [y 5]
      [z (* 2 x)])
  (+ z y))
```

But this fails to work. From a syntactic standpoint there's no way to convert this back to an equivalent lambda expression. And although the identifier `x` is bound in the list of binding expressions, the value of `x` is only available inside the body of the `let` expression.

There is, however, an alternative definition of `let` called `let*`. In this case the following would work.

```
> (let* ([x 4]
         [y 5]
         [z (* 2 x)])
    (+ z y))
13
```

The difference is that with `let*` the value of an identifier is available immediately after it's bound, whereas with `let` the identifier values are only available after *all* the identifiers are bound.

Here's another slight variation where `let` *does* work.

```
> (let ([x 4]
        [y 5])
    (let ([z (* 2 x)])
      (+ z y)))
13
```

In this case the second let is within the lexical environment of the first let (but as we've seen, let* more efficiently encodes this type of nested construct). Hence x is available for use in the expression (* 2 x).

Conditional Expressions: It's All About Choices

The ability of a computer to alter its execution path based on an input is an essential component of its architecture. Without this a computer cannot compute. In most programming languages this capability takes the form of something called a *conditional expression*, and in Racket it's expressed (in its most general form) as a cond expression.

Suppose you're given the task to write a function that returns a value that indicates whether a number is divisible by 3 only, divisible by 5 only, or divisible by both. One way to accomplish this is with the following code.

```
(define (div-3-5 n)
  (let ([div3 (= 0 (remainder n 3))]
        [div5 (= 0 (remainder n 5))])
    (cond [(and div3 div5) 'div-by-both]
          [div3 'div-by-3]
          [div5 'div-by-5]
          [else 'div-by-neither])))
```

The cond form contains a list of expressions. For each of these expressions, the first element contains some type of test, which if it evaluates to true, evaluates the second element and returns its value. Note that in this example the test for divisibility by 3 and 5 must come first. Here are trial runs:

```
> (div-3-5 10)
'div-by-5

> (div-3-5 6)
'div-by-3

> (div-3-5 15)
'div-by-both

> (div-3-5 11)
'div-by-neither
```

A simplified version of cond is the if form. This form consists of a single test (the first subexpression) that returns its second argument (after it's evaluated) if the test evaluates to true; otherwise it evaluates and returns the third argument. This example simply tests whether a number is even or odd.

```
(define (parity n)
  (if (= 0 (remainder n 2)) 'even 'odd))
```

If we run some tests:

```
> (parity 5)
'odd
> (parity 4)
'even
```

Both `cond` and `if` are expressions that return values. There are occasions where one simply wants to conditionally execute some sequence of steps if a condition is true or false. This usually involves cases where some side effect like printing a value is desired and returning a result is not required. For this purpose, Racket provides `when` and `unless`. If the conditional expression evaluates to true, `when` evaluates all the expressions in its body; otherwise it does nothing.

```
> (when (> 5 4)
    (displayln 'a)
    (displayln 'b))
a
b

> (when (< 5 4) ; doesn't generate output
    (displayln 'a)
    (displayln 'b))
```

The `unless` form behaves in exactly the same way as `when`; the difference is that `unless` evaluates its body if the conditional expression is not true.

```
> (unless (> 5 4) ; doesn't generate output
    (displayln 'a)
    (displayln 'b))

> (unless (< 5 4)
    (displayln 'a)
    (displayln 'b))
a
b
```

I'm Feeling a Bit Loopy!

Loops (or iteration) are the bread and butter of any programming language. With the discussion of loops, invariably the topic of *mutability* comes up. Mutability of course implies change. Examples of mutability are assigning values to variables (or worse, changing a value embedded in a data structure such as a vector). A function is said to be *pure* if no mutations (or side effects, like printing out a value or writing to a file—also forms of mutation) occur within the body of a function. Mutations are generally to be avoided if possible. Some languages, such as Haskell, go out of their way to avoid this type of mischief. A Haskell programmer would rather walk barefoot through a bed of glowing, hot coals than write an impure function.

There are many good reasons to prefer pure functions, such as something called referential transparency (this mouthful simply means the ability to reason about the behavior of your program). We won't be quite so persnickety and will make judicious use of mutation and impure functions where necessary.

Suppose you're given the task of defining a function to add the first *n* positive integers. If you're familiar with a language like Python (an excellent language in its own right), you might implement it as follows.

```
def sum(n):
    s = 0
    while n > 0:
     ➊ s = s + n
     ➋ n = n - 1
    return s
```

This is a perfectly good function (and a fairly benign example of using mutable variables) to generate the desired sum, but notice both the variables `s` and `n` are modified ➊ ➋. While there's nothing inherently wrong with this, these assignments make the implementation of the function `sum` impure.

Purity

Before we get down and dirty, let's begin by seeing how we can implement looping using only pure functions. *Recursion* is the custom when it comes to looping or iteration in Racket (and all functional programming languages). A recursive function is just a function defined in terms of itself. Here's a pure (and simple) recursive program to return the sum of the first *n* positive integers.

```
(define (sum n)
➊ (if (= 0 n) 0
   ➋ (+ n (sum (- n 1)))))
```

As you can see, we first test whether `n` has reached 0 ➊, and if so we simply return the value 0. Otherwise, we take the current value of `n` and *recursively* add to it the `sum` of all the numbers less than `n` ➋. For the mathematically inclined, this is somewhat reminiscent of how a proof by mathematical induction works where we have a base case ➊ and the inductive part of the proof ➋.

Let's test it out.

```
> (sum 100)
5050
```

There's a potential problem with the example we have just seen. The problem is that every time a recursive call is made, Racket must keep track of where it is in the code so that it can return to the proper place. Let's take a deeper look at this function.

```
(define (sum n)
  (if (= 0 n) 0
  ❶ (+ n (sum (- n 1)))))
```

When the recursive call to sum is made ❶, there's still an addition remaining to be done after the recursive call returns. The system must then remember where it was when the recursive call was made so that it can pick up where it left off when the recursive call returns. This isn't a problem for functions that don't have to nest very deeply, but for large depths of recursion, the computer can run out of space and fail in a dramatic fashion.

Racket (and virtually all Scheme variants) implement something called *tail call optimization* (the Racket community says this is simply the proper way to handle tail calls rather than an optimization, but *tail call optimization* is generally used elsewhere). What this means is that if a recursive call is the very last call being made, there's no need to remember where to return to since there are no further computations to be made within the function. Such functions in effect behave as a simple iterative loop. This is a basic paradigm for performing looping computations in the Lisp family of languages. You do, however, have to construct your functions in such a way as to take advantage of this feature. We can rewrite the summing function as follows.

```
(define (sum n)
  (define (s n acc)
 ❶ (if (= 0 n) acc
     ❷ (s (- n 1) (+ acc n))))
  (s n 0))
```

Notice that sum now has a local function called s that takes an additional argument called acc. Also notice that s calls itself recursively ❷, but it's the last call in the local function; hence tail call optimization takes place. This all works because acc accumulates the sum and passes it along as it goes. When it reaches the final nested call ❶, the accumulated value is returned.

Another way to do this is with a named let form as shown here.

```
(define (sum n)
  (let loop ([n n] [acc 0])
    (if (= 0 n) acc
        (loop (- n 1) (+ acc n)))))
```

The named let form, similar to the normal let, has a section where local variables are initialized. The expression [n n] may at first appear puzzling, but what it means is that the first n, which is local to the let, is initialized with the n that the sum function is called with. Unlike define, which simply binds an identifier with a function body, the named let binds the identifier (in this case loop), evaluates the body, and returns the value resulting from calling the function with the initialized parameter list. In this example the function is called recursively (which is the normal use case for a named let) as indicated by the last line in the code. This is a simple illustration of a side-effect-free looping construct favored by the Lisp community.

The Power of the Dark Side

Purity is good, as far as it goes. The problem is that staying pure takes a lot of work (especially in real life). It's time to take a closer look at the dreaded set! form. Note that an exclamation point at the end of any built-in Racket identifier is likely there as a warning that it's going to do something impure, like modify the program state in some fashion. A programming style that uses statements to change a program's state is said to use *imperative programming*. In any case, set! reassigns a value to a previously bound identifier. Let's revisit the Python sum function we saw a bit earlier. The equivalent Racket version is given below.

```
(define (sum n)
  (let ([s 0])   ; initialize s to zero
    (do ()       ; an optional initializer statement can go here
      ((< n 1))  ; do until this becomes true
      (set! s (+ s n))
      (set! n (- n 1)))
    s))
```

Racket doesn't actually have a while statement (this has to do with the expectation within the Lisp community that recursion *should* be the go-to method for recursion). The Racket do form functions as a do-until.

If you're familiar with the C family of programming languages, then you will see that the full form of the do statement actually functions much like the C for statement. One way to sum the first *n* integers in C would be as follows:

```
int sum(int n)
{
  int s = 0;
  for (i=1; i<= n; i++) // initialize i=1, set i = i+1 at each iteration
                        // do while i<= n
  {
      s = s + i;
  }
  return s;             // return s
}
```

Here's the Racket equivalent:

```
(define (sum n)
❶ (let ([s 0])
  ❷ (do ([i 1 (add1 i)])   ; initialize i=1, set i = i+1 at each iteration
```

```
❸ ((> i n) s)          ; do until i>n, then return s
❹ (set! s (+ s i)))))
```

In the above code we first initialize the local variable s (which holds our sum) to 0 ❶. The first argument to do ❷ initializes i (i is local to the do form) to 1 and specifies that i is to be incremented by 1 at each iteration of the loop. The second argument ❸ tests whether i has reached the target value and if so returns the current value of s. The last line ❹ is where the sum is actually computed by increasing the value of s with the current value of i via the set! statement.

The value of forms such as do with the set! statement is that many algorithms are naturally stated in a step-by-step fashion with variables mutated by equivalents to the set! statement. This helps to avoid the mental gymnastics needed to convert such constructs to pure recursive functions.

In the next section, we examine the for family of looping variants. Here we will see that Racket's for form provides a great deal of flexibility in how to manage loops.

The for Family

Racket provides the for form along with a large family of for variants that should satisfy most of your iteration needs.

A Stream of Values

Before we dive into for, let's take a look at a couple of Racket forms that are often used in conjunction with for: in-range and in-naturals. These functions return something we haven't seen before called a *stream*. A stream is an object that's sort of like a list, but whereas a list returns all its values at once, a stream only returns a value when requested. This is basically a form of *lazy evaluation*, where a value is not provided until asked for. For example, (in-range 10) will return a stream of 10 values starting with 0 and ending with 9. Here are some examples of in-range in action.

```
> (define digits (in-range 10))
> (stream-first digits)
0

> (stream-first (stream-rest digits))
1

> (stream-ref digits 5)
5
```

In the code above, (in-range 10) defines a sequence of values 0, 1, . . . , 9, but digits doesn't actually contain these digits. It basically just contains a specification that will allow it to return the numbers at some later time. When (stream-first digits) is executed, digits gives the first available value, which in this case is the number 0. Then (stream-rest digits) returns the stream containing the digits after the first, so that (stream-first (stream-rest digits)) returns the number 1. Finally, stream-ref returns the *i*-th value in the stream, which in this case is 5.

The function in-naturals works like in-range, but instead of returning a specific number of values, in-naturals returns an infinite number of values.

```
> (define naturals (in-naturals))
> (stream-first naturals)
0

> (stream-first (stream-rest naturals))
1

> (stream-ref naturals 1000)
1000
```

How the stream concept is useful will become clearer as we see it used within some for examples. We'll also met some useful additional arguments for in-range.

for in the Flesh

Here's an example of for in its most basic form. The goal is to print each character of the string "Hello" on a separate line.

```
> (let* ([h "Hello"]
       ❶ [l (string-length h)])
    ❷ (for ([i (in-range l)])
      ❸ (display (string-ref h i))
       (newline)))
H
e
l
l
o
```

We capture the string-length ❶ and use this length with the in-range function ❷. for then uses the resulting stream of values to populate the identifier i, which is used in the body of the for form to extract and display the characters ❸. In the prior section it was pointed out that in-range produces a sequence of values, but it turns out that in the context of a for statement, a positive integer can also produce a stream as the following example illustrates.

```
> (for ([i 5]) (display i))
01234
```

The for form is quite forgiving when it comes to the type of arguments that it accepts. It turns out that there's a much simpler way to achieve our goal.

```
> (for ([c "Hello"])
    (display c)
    (newline))
H
e
l
l
o
```

Instead of a stream of indexes, we have simply provided the string itself. As we'll see, `for` will accept many built-in data types that consist of multiple values, like lists, vectors, and sets. These data types can also be converted to streams (for example, by `in-list`, `in-vector`, and so on), which in some cases can provide better performance when used with `for`. All expressions that provide values to the identifier that `for` uses to iterate over are called *sequence expressions*.

It's time to see how we can make use of the mysterious `in-naturals` form introduced above.

```
> (define (list-chars str)
    (for ([c str]
          [i (in-naturals)])
      (printf "~a: ~a\n" i c)))

> (list-chars "Hello")
0: H
1: e
2: l
3: l
4: o
```

The `for` form inside the `list-chars` function now has *two* sequence expressions. Such sequence expressions are evaluated in parallel until one of the expressions runs out of values. That is why the `for` expression eventually terminates, even though `in-naturals` provides an infinite number of values.

There is, in fact, a version of `for` that *does not* evaluate its sequence expressions in parallel: it's called `for*`. This version of `for` evaluates its sequence expressions in a nested fashion as the following example illustrates.

```
> (for* ([i (in-range 2 7 4)]
         [j (in-range 1 4)])
    (display (list i j (* i j)))
    (newline))
(2 1 2)
(2 2 4)
(2 3 6)
(6 1 6)
```

```
(6 2 12)
(6 3 18)
```

In this example we also illustrate the additional optional arguments that `in-range` can take. The sequence expression `(in-range 2 7 4)` will result in a stream that starts with the number 2, and increment that value by 4 with each iteration. The iteration will stop once the streamed value reaches one less than 7. So in this expression, `i` is bound to 2 and 6. The expression `(in-range 1 4)` does not specify a step value, so the default step size of 1 is used. This results in `j` being bound to 1, 2, and 3.

Ultimately, `for*` takes every possible combination of `i` values and `j` values to form the output shown.

Can You Comprehend This?

There is a type of notation in mathematics called set-builder notation. An example of set-builder notation is the expression $\{x^2 \mid x \in \mathbb{N}, x \leq 10\}$. This is just the set of squares of all the natural numbers between 0 and 10. Racket provides a natural (pun intended) extension of this idea in the form of something called a *list comprehension*. A direct translation of that mathematical expression in Racket would appear as follows.

```
> (for/list ([x (in-naturals)] #:break (> x 10)) (sqr x))
'(0 1 4 9 16 25 36 49 64 81 100)
```

The `#:break` keyword is used to terminate the stream generated by `in-naturals` once all the desired values have been produced. Another way to do this, without having to resort to using `#:break`, would be with `in-range`.

```
> (for/list ([x (in-range 11)]) (sqr x))
'(0 1 4 9 16 25 36 49 64 81 100)
```

If you only wanted the squares of even numbers, you could do it this way:

```
> (for/list ([x (in-range 11)] #:when (even? x)) (sqr x))
'(0 4 16 36 64 100)
```

This time the `#:when` keyword was brought into play to provide a condition to filter the values used to generate the list.

An important difference of `for/list` over `for` is that `for/list` does not produce any side effects and is therefore a pure form, whereas `for` is expressly for the purpose of producing side effects.

More Fun with for

Both `for` and `for/list` share the same keyword parameters. Suppose we wanted to print a list of squares, but don't particularly like the number 5. Here's how it could be done.

```
> (for ([n (in-range 1 10)] #:unless (= n 5))
    (printf "~a: ~a\n" n (sqr n)))
1: 1
2: 4
3: 9
4: 16
6: 36
7: 49
8: 64
9: 81
```

By using `#:unless` we've produced an output for all values, $1 \leq n < 10$, unless $n = 5$.

Sometimes it's desirable to test a list of values to see if they all meet some particular criteria. Mathematicians use a fancy notation to designate this called the universal quantifier, which looks like this $\forall$ and means "for all." An example is the expression $\forall x \in \{2, 4, 6\}, x \bmod 2 = 0$, which is literally interpreted as "for all x in the set $\{2, 4, 6\}$, the remainder of x after dividing by 2 is 0." This just says that the numbers 2, 4, and 6 are even. The Racket version of "for all" is `for/and`.

Feed the `for/and` form a list of values and a Boolean expression to evaluate the values. If each value evaluates to true, the entire `for/and` expression returns true; otherwise it returns false. Let's have a go at it.

```
> (for/and ([x '(2 4 6)]) (even? x))
#t

> (for/and ([x '(2 4 5 6)]) (even? x))
#f
```

Like `for`, `for/and` can handle multiple sequence expressions. In this case, the values in each sequence are compared in parallel.

```
> (for/and ([x '(2 4 5 6)]
            [y #(3 5 9 8)])
    (< x y))
#t
> (for/and ([x '(2 6 5 6)]
            [y #(3 5 9 8)])
    (< x y))
#f
```

Closely related to `for/and` is `for/or`. Not to be outdone, mathematicians have a notation for this as well: it's called the existential quantifier, $\exists$. For example, they express the fact that there *exists* a number in the set $\{2, 7, 4, 6\}$ greater than 5 with the expression $\exists x \in \{2, 7, 4, 6\}, x > 5$.

```
> (for/or ([x '(2 7 4 6)]) (> x 5))
#t

> (for/or ([x '(2 1 4 5)]) (> x 5))
#f
```

Suppose now that you not only want to know whether a list contains a value that meets a certain criterion, but you want to extract the first value that meets the criterion. This is a job for `for/first`:

```
> (for/first ([x '(2 1 4 6 7 1)] #:when (> x 5)) x)
6

> (for/first ([x '(2 1 4 5 2)] #:when (> x 5)) x)
#f
```

The last example demonstrates that if there is no value that meets the criterion, `for/first` returns false.

Correspondingly, if you want the last value, you can use `for/last`:

```
> (for/last ([x '(2 1 4 6 7 1)] #:when (> x 5)) x)
7
```

The `for` family of functions is fertile ground for exploring parallels between mathematical notation and Racket forms. Here is yet another example. To indicate the sum of the squares of the integers from 1 to 10, the following notation would be employed:

$$S = \sum_{i=1}^{10} i^2$$

The equivalent Racket expression is:

```
> (for/sum ([i (in-range 1 11)]) (sqr i))
385
```

The equivalent mathematical expression for products is

$$p = \prod_{i=1}^{10} i^2$$

which in Racket becomes

```
> (for/product ([i (in-range 1 11)]) (sqr i))
13168189440000
```

Most of the `for` forms discussed above come in a starred version (for example `for*/list`, `for*/and`, `for*/or`, and so on). Each of these works by evaluating their sequence expressions in a nested fashion as described for `for*`.

Time for Some Closure

Suppose you had $100 in the bank and wanted to explore the effects of compounding with various interest rates. If you're not familiar with how compound interest works (and you very well should be), it works as follows: if you have n_0 in a bank account that pays i periodic interest, at the end of the period you would have this:

$$n_1 = n_0 + n_0 i = n_0(1 + i)$$

Using your $100 deposit as an example, if your bank pays 4 percent (i = 0.04) interest per period (good luck getting that rate at a bank nowadays), you would have the following at the end of the period:

$$100 + 100 \cdot 4\% = 100(1 + 0.04) = 104$$

One way to do this is to create a function that automatically updates the balance after applying the interest rate. A clever way to compute this in Racket is with something called a *closure*, which we use in the following function:

```
(define (make-comp bal int)
  (let ([rate (add1 (/ int 100.0))])
❶ (λ () (set! bal (* bal rate))  (round bal))))
```

Notice that this function actually returns another function—the lambda expression (λ . . .) ❶—and that the lambda expression contains variables from the defining scope. We shall explain how this works shortly.

In the code above, we've defined a function called `make-comp` which takes two arguments: the starting balance and the interest rate percentage. The `rate` variable is initialized to (1 + i). Rather than return a number, this function actually returns another function. The returned function is designed in such a way that every time it's called (without arguments) it updates the balance by applying the interest and returns the new balance. You might think that once `make-comp` returns the lambda expression, the variables `bal` and `rate` would be undefined, but not so with closures. The lambda expression is said to *capture* the variables `bal` and `rate`, which are available within the lexical environment where the lambda expression is defined. The fact that the returned function contains the variables `bal` and `rate` (which are defined outside of the function) is what makes it a closure.

Let's try this out and see what happens.

```
> (define bal (make-comp 100 4))

> (bal)
104.0

> (bal)
108.0
```

```
> (bal)
112.0

> (bal)
117.0
```

As you can see, the balance is updated appropriately.

Another use for closures is in a technique called *memoization*. What this means is that we store prior computed values and if a value has already been computed, return the remembered value; otherwise go ahead and compute the value and save it for when it's needed again. This is valuable in scenarios where a function may be called repeatedly with arguments that have already been computed.

To facilitate this capability, something called a *hash table* or dictionary is typically used. A hash table is a mutable set of key-value pairs. A hash table is constructed with the function `make-hash`. Items can be stored to the hash table via `hash-set!` and retrieved from the table with `hash-ref`. We test whether the table already contains a key with `hash-has-key?`.

The standard definition for the factorial function is $n! = n(n - 1)!$. The obvious way to implement this in Racket is with the following.

```
(define (fact n)
  (if ( = 0 n) 1
      (* n (fact (- n 1)))))
```

This works, but every time you call `(fact 100)`, Racket has to perform 100 computations. With memoization, executing `(fact 100)` still requires 100 computations *the first time*. But the next time you call `(fact 100)` (or call `fact` for any value less than 100), Racket only has to look up the value in the hash table, which happens in a single step. Here's the implementation.

```
(define fact
  (let ([h (make-hash)]) ; hash table to contain memoized values
 ❶ (define (fact n)
      (cond [(= n 0) 1]
          ❷ [(hash-has-key? h n) (hash-ref h n)]
            [else
            ❸ (let ([f (* n (fact (- n 1)))])
              ❹ (hash-set! h n f)
                f)]))
 ❺ fact))
```

It's important to note that the outer `fact` function actually returns the inner `fact` function ❶. This is ultimately what gets executed when we call `fact 100`. It's this inner `fact` function, which captures the hash table, that constitutes the closure. First, it checks to see whether the argument to `fact` is one that is already computed ❷ and if so, returns the saved value. We still have to compute the value if it hasn't been computed yet ❸, but then we save it in case it's needed later ❹. The local fact function is returned as the value of the global fact function (sorry about using the same name twice).

Applications

Having introduced the basic programming constructs available in Racket, let's take a look at some applications spanning computer science, mathematics, and recreational puzzles.

I Don't Have a Queue

In this section we touch on Racket's *object-oriented programming* capability. Objects are like a deluxe version of the structures we met in Chapter 1.

Imagine early morning at a small-town bank with a single teller. The bank has just opened, and the teller is still trying to get set up, but a customer, Tom, has already arrived and is waiting at the window. Shortly, two other customers show up: Dick and Harry. The teller finally waits on Tom, then Dick and Harry in that order. This situation is a classic example of a *queue*. Formally, a queue is a first-in, first-out (FIFO) data structure. Racket comes with a built-in queue (several, in fact), but let's explore building one from scratch.

We can model a queue with a list. For example the line of folks waiting to see the teller can be represented by a single list: `(define q (list 'tom 'dick 'harry))`. But there's a problem. It's clearly easy to remove Tom from the head of the list and get the remainder of the list by using `car` (or `first`) and `cdr` (or `rest`):

```
> (car q)
'tom

> (set! q (cdr q))
> q
'(dick harry)
```

But what happens when Sue comes along? We could do the following:

```
> (set! q (append q (list 'sue)))
> q
'(dick harry sue)
```

But consider what happens if the list is very long, say 10,000 elements. The `append` function will create an entire new list containing all the elements from q and the one additional value `'sue`. One way to do this efficiently is to maintain a pointer to the last element in the list and instead of creating a new list change the `cdr` of the last node of the list to point to the list `(list 'sue)` (see Figure 3-2). About now alarm bells should be going off in your head. You should have an uneasy feeling that modifying a list structure is somehow wrong. And you'd be right. It's not even possible to do this with the normal

Racket list structure since the car and cdr cells in a list pair are immutable and cannot be changed.

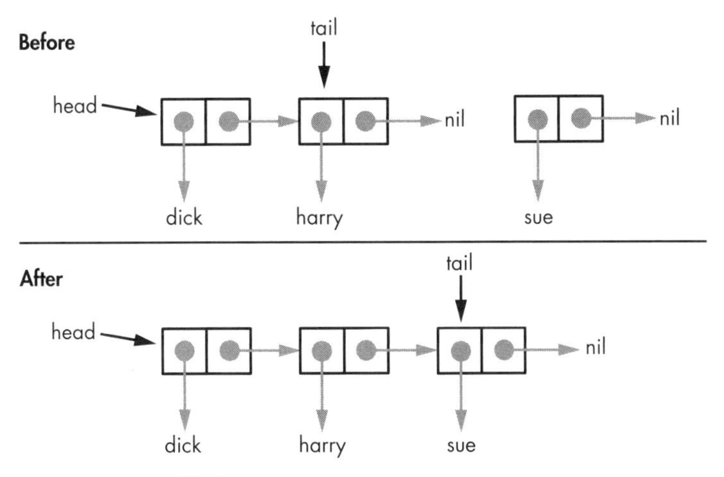

Figure 3-2: Mutable list

The traditional version of Scheme allow the elements of a cons node to be modified via `set-car!` and `set-cdr!` methods. Since these aren't defined in Racket, Racket guarantees that any identifier bound to a Racket list will have the same value for the life of the program.

There are still valid reasons why this capability may be needed. As we've seen, this functionality is needed for queues to ensure efficient operation. To accommodate this need, Racket provides a mutable cons cell that can be created with the `mcons` function. Each component of the mutable cons cell can be modified with `set-mcar!` and `set-mcdr!`. The functions `mcar` and `mcdr` are the corresponding accessor functions.

The reason modifying a list structure is bad is because if some other identifier is bound to the list, it will now have the modified list as its value, and maybe that's not what was intended. Observe the following.

```
> (define a (mcons 'apple 'orange))
> (define b a)
> a
(mcons 'apple 'orange)
> b
(mcons 'apple 'orange)

> (set-mcdr! a 'banana)
> a
(mcons 'apple 'banana)
> b
(mcons 'apple 'banana)
```

Although we only *seemed* to be changing the value of `a`, we also changed the value of `b`.

To avoid this potentially disastrous situation, we'll *encapsulate* the list in such a way that the list itself is not accessible, but we'll still be able to remove elements from the front of the list and add elements to the end of the list to implement our queue. Encapsulation is a fundamental component of object-oriented programming. We'll dive right in by creating a class that contains all the functionality we need to implement our queue:

```
❶ (define queue%

 ❷ (class object%

  ❸ (init [queue-list '()])

  ❹ (define head '{})
    (define tail '{})

  ❺ (super-new)

  ❻ (define/public (enqueue val)
      (let ([t (mcons val '())])
        (if (null? head)
            (begin
              (set! head t)
              (set! tail t))
            (begin
              (set-mcdr! tail t)
              (set! tail t)))))

  ❼ (define/public (dequeue)
      (if (null? head) (error "Queue is empty.")
          (let ([val (mcar head)])
          ❽ (set! head (mcdr head))
            (when (null? head) (set! tail '()))
            val)))

    (define/public (print-queue)
      (define (prt rest)
        (if (null? rest)
            (newline)
            (let ([h (mcar rest)]
                  [t (mcdr rest)])
              (printf "~a " h)
              (prt t))))
        (prt head))

  ❾ (for ([v queue-list]) (enqueue v))))
```

Our class name is queue% (note that, by convention, Racket class names end with %). We begin with the class definition ❶. All classes must inherit from some parent class. In this case we're using the built-in class object% ❷. Once we've specified the class name and parent class, we specify the initialization parameters for the class ❸. This class takes a single, optional list argument. If supplied, this list is used to initialize the queue ❾. Our class uses head and tail pointer identifiers, which we have to define ❹. Within the body of a class, define statements are not accessible from outside the class. This means that there is no way for the values of head or tail to be bound to an identifier outside of the class.

After a required call to the super class (in this case object%) ❺, we get into the real meat of this class: its methods. First we define a *public* class method called enqueue ❻. Public methods are accessible from outside the class. This method takes a single value, which is added to the end of the queue in a manner similar to our apple and banana example. If the queue is empty, then it initializes the head and tail identifiers with the mutable cons cell t.

The dequeue method ❼ returns the value at the head of the queue, but generates an error if the queue is empty. The head pointer is updated to point to the next value in the queue ❽.

To see all the values in the queue, we've also defined the method print-queue.

Let's see it in action.

```
> (define queue (new queue% [queue-list '(tom dick harry)]))

> (send queue dequeue)
'tom

> (send queue enqueue 'sue)
> (send queue print-queue)
dick harry sue

> (send queue dequeue)
'dick

> (send queue dequeue)
'harry

> (send queue dequeue)
'sue

> (send queue dequeue)
. . Queue is empty.
```

Class objects are created with the new form. This form includes the class name and any parameters defined by the init form in the class definition (see the class definition code ❸).

Unlike normal Racket functions and methods, an object method must be invoked with a `send` form. The `send` identifier is followed by the object name (`queue`), the method name, and any arguments for the method.

This example was just meant to expose the basics of Racket's object-oriented capabilities, but we'll be seeing much more of Racket's object prowess in the remainder of the text.

The Tower of Hanoi

The Tower of Hanoi is a puzzle that consists of three pegs embedded in a board, along with eight circular discs, each with a hole in the center. No two discs are the same size, and they are arranged on one of the pegs so that the largest is on the bottom and the rest are arranged such that a smaller disc is always immediately above a larger disc (See Figure 3-3).

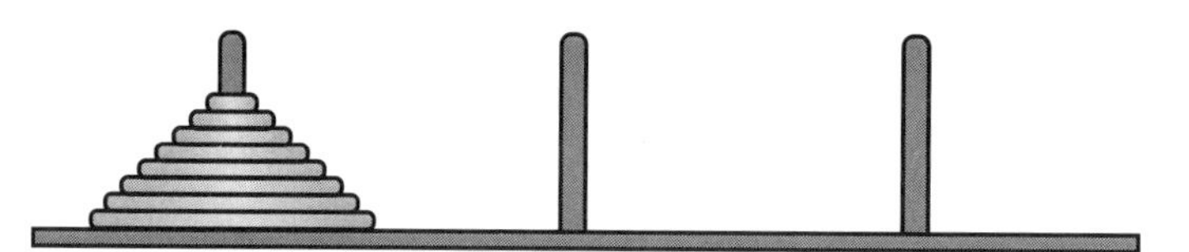

Figure 3-3: The Tower of Hanoi

W. W. Rouse Ball tells the following entertaining story about how this puzzle came about (see [3] and [8]).

> In the great temple at Benares beneath the dome which marks the center of the world, rests a brass plate in which are fixed three diamond needles, each a cubit high and as thick as the body of a bee. On one of these needles, at the creation, God placed sixty-four disks of pure gold, the largest disc resting on the brass plate and the others getting smaller and smaller up to the top one. This is the tower of Brahma. Day and night unceasingly, the priest on duty transfers the disks from one diamond needle to another, according to the fixed and immutable laws of Brahmah, which require that the priest must move only one disk at a time, and he must place these discs on needles so that there never is a smaller disc below a larger one. When all the sixty-four discs shall have been thus transferred from the needle on which, at the creation, God placed them, to one of the other needles, tower, temple, and Brahmans alike will crumble into dust, and with a thunderclap the world will vanish.

This would take 2^{64} - 1 moves. Let's see how much time we have left until the world comes to an end. We assume one move can be made each second.

```
> (define moves (- (expt 2 64) 1))
> moves
18446744073709551615

> (define seconds-in-a-year (* 60 60 24 365.25))
> seconds-in-a-year
```

```
31557600.0

> (/ moves seconds-in-a-year)
584542046090.6263
```

This last number is about 5.84×10^{11} years. The universe is currently estimated to be a shade under 14×10^9 years old. If the priests started moving disks at the beginning of the universe, there would be about 570 billion years left, so you probably have at least enough time to finish reading this book.

As interesting as this is, our main objective is to use Racket to show how to actually perform the moves. We'll of course begin with a more modest number of disks, so let's start with just one disk. We'll number the pegs 0, 1, and 2. Suppose our goal is to move the disks from peg 0 to peg 2. With only one disk, we just move the disk from peg 0 to peg 2. If we have $n > 1$ disks, we designate the peg we're moving all the disks from as *f*, the peg we are moving to as *t* and the remaining peg we designate as *u*. The steps to solve the puzzle can be stated thusly:

1. Move n - 1 disks from *f* to *u*.
2. Move a single disk from *f* to *t*.
3. Move n - 1 disks from *u* to *t*.

While simple, this process is sufficient to solve the puzzle. Steps 1 and 3 imply the use of recursion. Here is the Racket code that implements these steps.

```
❶ (define (hanoi n f t)
  ❷ (if (= 1 n) (list (list f t))        ; only a single disk to move
     ❸ (let* ([u (- 3 (+ f t))]          ; determine unused peg
            ❹ [m1 (hanoi (sub1 n) f u)] ; move n-1 disks from f to u
            ❺ [m2 (list f t)]           ; move single disk from f to t
            ❻ [m3 (hanoi (sub1 n) u t)]); move disks from u to t
       ❼ (append m1 (cons m2 m3)))))
```

We pass hanoi the number of disks, the peg to move them from, and the peg to move to. Then we compute the moves required to implement steps one ❸, two ❹, and three ❺. Can you see why the let expression ❸ determines the unused peg? (Hint: think of the possible combinations. For example if $f = 1$ and $t = 2$, the let expression ❸ would give $u = 3 - (1 + 2) = 0$, the unused peg number.) The hanoi function returns a list of moves ❻. Each element of the list is a list of two elements that designate the peg to move from and the peg to move to. Here's an example of the output for three disks:

```
> (hanoi 3 0 2)
'((0 2) (0 1) (2 1) (0 2) (1 0) (1 2) (0 2))
```

Note that we have $2^3 - 1 = 7$ moves.

As can be seen from the comments in the code, the `hanoi` function is essentially a direct translation of the three-step solution process given earlier. Further, it provides a practical application of recursion where the function calls itself with a simpler version of the problem.

Fibonacci and Friends

The *Fibonacci sequence* of numbers is defined as

$$0, 1, 1, 2, 3, 5, 8, 13, 21, 34, \ldots$$

where the next term in the sequence is always the sum of the two preceding terms. In some cases the initial zero is not considered part of the sequence. This sequence has a ton of interesting properties. We will only touch on a few of them here.

Some Interesting Properties

One interesting property of the Fibonacci sequence is that it's always possible to create a rectangle tiled with squares whose sides have lengths generated by the sequence, as seen in Figure 3-4. We'll see how to generate this tiling in Chapter 4.

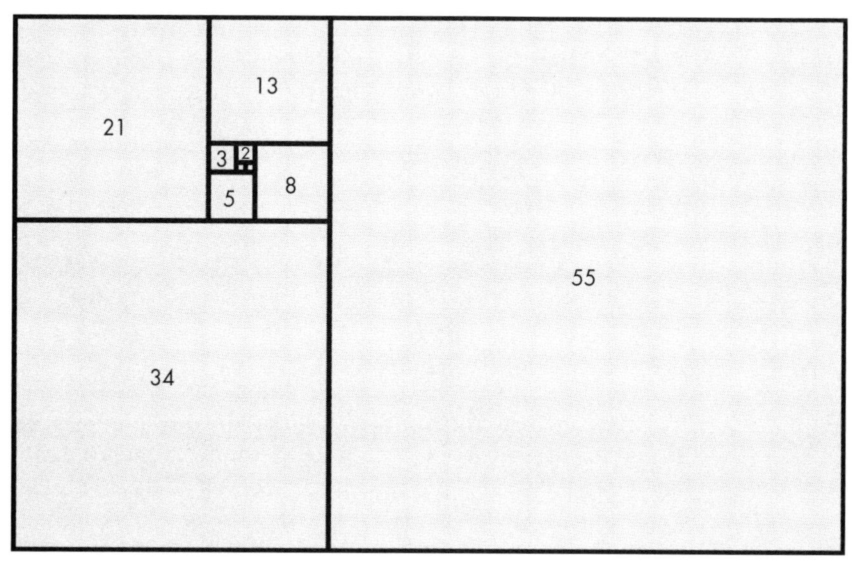

Figure 3-4: Fibonacci tiling

Johannes Kepler pointed out that the ratio of consecutive Fibonacci numbers approaches a particular number, designated by ϕ, which is known as the *golden ratio*:

$$\lim_{n\to\infty} \frac{F_{n+1}}{F_n} = \phi = \frac{1+\sqrt{5}}{2} \approx 1.6180339887\ldots$$

If you're not familiar with that lim $n \to \infty$ business, it just means this is what you get when n gets bigger and bigger.

The number ϕ has many interesting properties as well. One example is the *golden spiral*. A golden spiral is a logarithmic spiral whose growth factor is ϕ, which means that it gets wider (or further from its origin) by a factor of ϕ for every quarter-turn. A golden spiral with initial radius 1 has the following polar equation:

$$r = \phi^{\theta \frac{2}{\pi}} \tag{3.1}$$

A plot of the golden spiral is shown in Figure 3-5. We'll show how this plot was produced in Chapter 4.

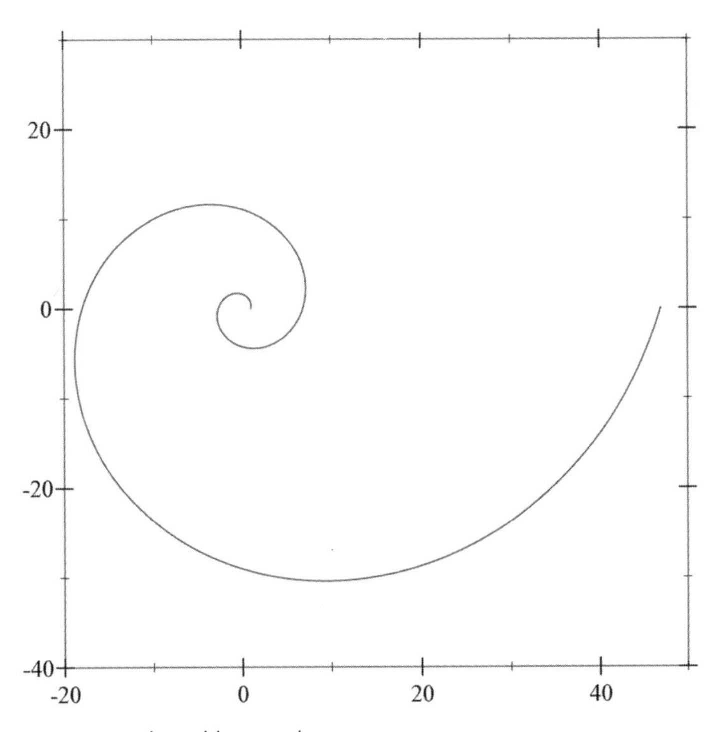

Figure 3-5: The golden spiral

Figure 3-6 illustrates an approximation of the golden spiral created by drawing circular arcs connecting the opposite corners of squares in the Fibonacci tiling (in Chapter 4 we'll see how to superimpose this spiral onto a Fibonacci tiling).

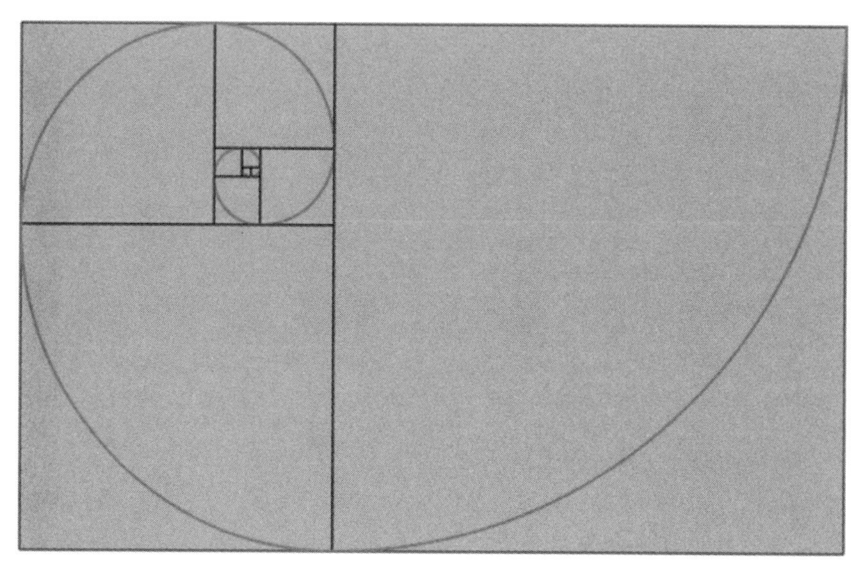

Figure 3-6: A golden spiral approximation

While these two versions of the golden spiral appear quite similar, mathematically they're quite different. This has to do with a concept called *curvature*. This has a precise mathematical definition, but for now, just think of it as the curviness of the path. The tighter the curve, the larger the curviness. The path described by Equation (3.1) has continuous curvature, while the Fibonacci spiral has discontinuous curvature. Figure 3-7 demonstrates the distinct difference in curvature these two paths possess.

We will make use of these properties in the following sections and in Chapter 4.

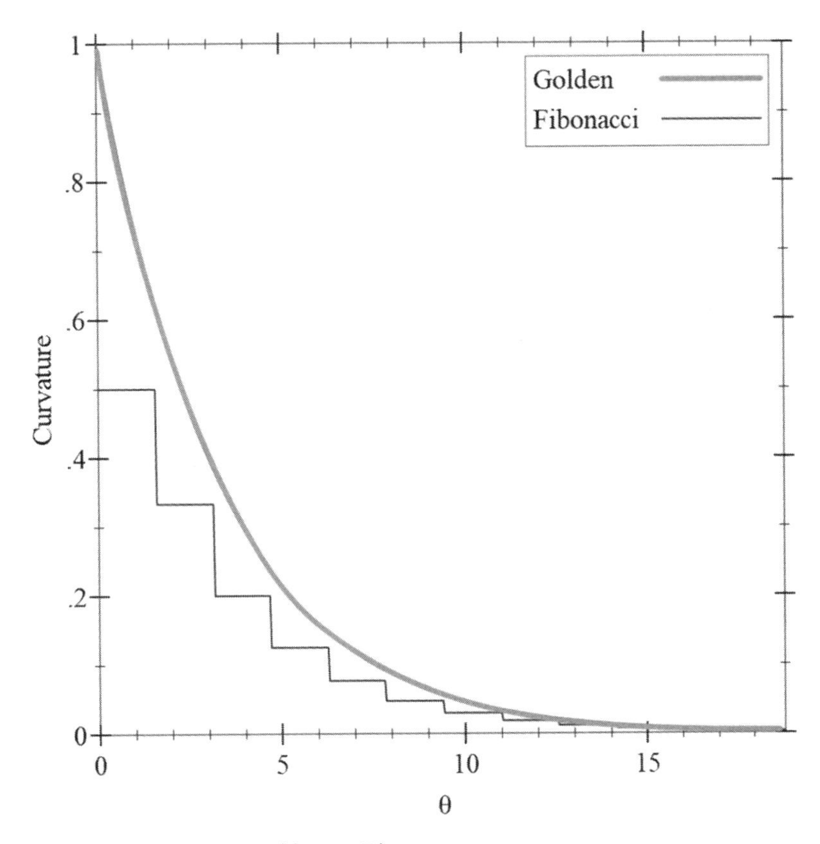

Figure 3-7: Curvature: golden vs. Fibonacci

Computing the Sequence

Mathematically the sequence F_n of Fibonacci numbers is defined by the recurrence relation:

$$F_n = F_{n-1} + F_{n-2}$$

In this section we'll explore three different methods of computing this sequence.

1. The No-brainer Approach. With the recurrence relation definition for the Fibonacci sequence, our first version practically writes itself. It's literally an exact translation from the definition to a Racket function.

```
(define (F n)
  (if (<= n 1) n
    (+ (F (- n 1)) (F (- n 2)))))
```

This code has the virtue of being extremely clear and simple. The only problem with this code is that it's terribly inefficient. The two nested calls cause the same value to be computed over and over. The end result is that the amount of computation grows exponentially with the size of n.

2. Efficiency Is King. Here we explore an ingenious method presented in the computer science classic *Structure and Interpretation of Computer Programs* [2]. The idea is to use a pair of integers initialized such that $a = F_1 = 1$ and $b = F_0 = 0$, and repeatedly apply the transformations:

$$a \leftarrow a + b$$
$$b \leftarrow a$$

It can be shown that after applying these transformations n times, we'll have $a = F_{n+1}$ and $b = F_n$. The proof is not difficult, and I've left it as an exercise for you. Here's the code to implement this solution:

```
(define (F n)
  (define (f a b c)
    (if (= c 0) b
        (f (+ a b) a (- c 1))))
  (f 1 0 n))
```

Due to tail call optimization, `f` recursively calls itself without the need to keep track of a continuation point. This works as an iterative process and only grows linearly with the size of n.

3. Memory Serves. In this version we use the memoization technique introduced in "Time for Some Closure" on page 58. To facilitate this, the code below uses a *hash table*. Recall that a hash table is a mutable set of key-value pairs, and it's constructed with the function `make-hash`. Items can be stored to the hash table via `hash-set!` and retrieved from the table with `hash-ref`. We test whether the table already contains a key with `hash-has-key?`.

```
(define F
  (let ([f (make-hash)]) ; hash table to contain memoized F values
    (define (fib n)
      (cond [(<= n 1) n]
            [(hash-has-key? f n) (hash-ref f n)]
            [else
              (let ([fn (+ (fib (- n 1)) (fib (- n 2)))])
                (hash-set! f n fn)
                fn)]))
    fib))
```

This code should be fairly easy to decipher. It's a straightforward application of memoization as seen in the `fact` example presented earlier.

And the Winner Is? It depends. You should definitely never use the first approach. Here's something to consider when comparing the second and the third: the second approach *always* requires n computations every time `F` is called for n. The third approach also requires n computations *the first time*

F is called for n. If you call it a second (or subsequent) time for n (or for any number less than n), it returns almost instantly since it simply has to look up the value in the hash table. There's a small space penalty for the third approach, but it's likely to be insignificant in most cases.

Binet's Formula. Before we leave the fascinating world of Fibonacci numbers and how to compute them, let's take a look at *Binet's Formula*:

$$F_n = \frac{\phi^n - \psi^n}{\phi - \psi} = \frac{\phi^n - \psi^n}{\sqrt{5}}$$

In this formula, the following is true:

$$\psi = \frac{1 - \sqrt{5}}{2} = 1 - \phi = -\frac{1}{\phi}$$

This formula provides us with yet another way of computing F_n. The following applies to all n:

$$\left| \frac{\psi^n}{\sqrt{5}} \right| < \frac{1}{2}$$

So the number F_n is the closest integer to $\frac{\psi^n}{\sqrt{5}}$. Therefore, if we round to the nearest integer, F_n can be computed by the following:

$$F_n = \left[\frac{\psi^n}{\sqrt{5}} \right]$$

The square brackets are used to designate the rounding function. In Racket, this becomes:

```
(define (F n)
  (let* ([phi (/ (add1 (sqrt 5)) 2)]
         [phi^n (expt phi n)])
    (round (/ phi^n (sqrt 5)))))
```

While Binet's formula is quite fast (since it does not require looping or recursion), the downside is that it only gives an approximate answer, where the other versions give an exact value.

Continued Fractions. The expression below is an example of a *continued fraction*. In this case, the fractional portion is repeated indefinitely. As we shall see, continued fractions have a surprising relationship to the Fibonacci sequence.

$$f = 1 + \cfrac{1}{1 + \cfrac{1}{1 + \cfrac{1}{1 + \cdots}}}$$

Since the fraction does repeat infinitely, we may make the following substitution.

$$f = 1 + \frac{1}{f}$$

This substitution simplifies to the quadratic equation:

$$f^2 - f - 1 = 0$$

That equation has a couple of solutions. This is true:

$$f = \frac{1 \pm \sqrt{5}}{2}$$

Or, these are true:

$$\phi = \frac{1 + \sqrt{5}}{2} \text{ and } \psi = \frac{1 - \sqrt{5}}{2}$$

The question remains: which of these is the right value for f? Since ψ is negative, the answer must be ϕ. Thus ...

$$\phi = 1 + \cfrac{1}{1 + \cfrac{1}{1 + \cfrac{1}{1 + \cdots}}}$$

Bet you didn't see that coming.

The Insurance Salesman Problem

This problem is adapted from Flannery's *In Code* [7]. It's an example of a problem that could be solved by hand, but we can take advantage of Racket to do some of the tedious calculations. The problem is stated as follows.

A door-to-door insurance salesman stops at a woman's house and the following dialog ensues:

> Salesman: How many children do you have?
>
> Woman: Three.
>
> Salesman: And what are their ages?
>
> Woman: Take a guess.
>
> Salesman: How about a hint?
>
> Woman: Okay, the product of their ages is 36 and all the ages are whole numbers.
>
> Salesman: That's not much to go on. Can you give me another hint?
>
> Woman: The sum of their ages is equal to the number on the house next door.
>
> The salesman immediately runs off, jumps over the fence, looks at the number on the house next door, scratches his head, and goes back to the woman.
>
> Salesman: Could you give me just one more hint?
>
> Woman: The oldest one plays the piano.

> The salesman thinks for a bit, does some calculations, and figures out the children's ages. What are they?

At first blush, the hints seem a bit incongruous. Let's take them one at a time. First, we know that the product of the three ages is 36. Here is a program that generates all the unique combinations of three positive integers that have a product of 36.

```
#lang racket
(require math/number-theory)

❶ (define triples '())
  (define (gen-triples d1)
 ❷ (let* ([q (/ 36 d1)]
          [divs (divisors q)])
   ❸ (define (try-div divs)
        (when (not (null? divs))
       ❹ (let* ([d2 (car divs)] [d3 (/ q d2)])
         ❺ (when (<= d3 d2 d1)
           ❻ (set! triples (cons (list d3 d2 d1) triples)))
            (try-div (cdr divs)))))
      (try-div divs)))

❼ (for ([d (divisors 36)]) (gen-triples d))

  triples
```

While this code will not win any awards for efficiency, it is relatively simple and it gets the job done. We first define the variable `triples` which will contain the list of generated triples ❶. The processing actually begins when we call `gen-triples` ❼ for each divisor of 36 (provided by the `divisors` function defined in the *math/number-theory* library). This function then defines the quotient `q` ❷ of the divisor `d1` into 36. Following this we generate a list of divisors of `q` (`divs`, which of course also divide 36). We now come to the function `try-div` ❸, which does the bulk of the work. Then we get the first divisor (`d2`) of `q` ❹ and generate the third divisor (`d3`) by dividing `q` by `d2`. These divisors (`d1`, `d2`, and `d3`) are tested to see whether a satisfactory triple is formed (that is to ensure uniqueness, we make sure that they form an ordered sequence ❺) . If so, it's added to the list of triples ❻. Testing other divisors resumes on the following line. Running this program produces the following sets of triples: {1,1,36}, {1,2,18}, {1,3,12}, {1,4,9}, {2,2,9}, {1,6,6}, {2,3,6}, {3,3,4}.

This alone, of course, does not allow the salesman to determine the ages of the children. The second hint is that the sum of the ages equals the number on the house next door. Again we make use of Racket to generate the required sums.

```
(for ([triple triples]) (printf "~a: ~a\n" triple (apply + triple)))
```

From this, we have the following.

$$
\begin{aligned}
1 + 1 + 36 &= 38 \\
1 + 2 + 18 &= 21 \\
1 + 3 + 12 &= 16 \\
1 + 4 + 9 &= 14 \\
2 + 2 + 9 &= 13 \\
1 + 6 + 6 &= 13 \\
2 + 3 + 6 &= 11 \\
3 + 3 + 4 &= 10
\end{aligned}
$$

After looking at the number of the house next door, the salesman still does not know the ages. This means the ages must have been one of the two sets of numbers that sum to 13 (otherwise he would have known which set to select). Since the woman said "The oldest *one* plays the piano," the only possibility is the set of ages $\{2, 2, 9\}$, since the set $\{1, 6, 6\}$ would imply *two* oldest.

Summary

In this chapter, we introduced Racket's basic programming constructs and applied them to a variety of problem domains. So far our explorations have been confined to getting output in a textual form. Next, we will see how to add some bling to our applications by generating some graphical output.

4

PLOTTING, DRAWING, AND A BIT OF SET THEORY

In this chapter, we'll explore how to exploit DrRacket's ability to display graphical information. In particular, we'll look at various ways of generating two-dimensional function plots as well as using some built-in graphics primitives (circles, rectangles, and so on) to create more-complex drawings. After introducing the basics, we'll look at a few extended applications, involving set theory, the golden spiral, and the game of Nim.

Plotting

All plotting functions will require the *plot* library, so be sure to execute the following before trying any of the examples in this section.

```
> (require plot)
```

With that out of the way, let's begin with something simple and familiar.

X-Y Plots

All two-dimensional plots use the plot function. This function takes either a single function or a list of functions, and generates a plot for each. Once a plot has been generated, if desired, it can be saved as a bitmap by right-clicking on the plot and clicking Save Image. . . from the pop-up menu.

The plot function accepts a number of keyword arguments (all of which are optional), but as a minimum, it's probably a good idea to specify the plot limits since DrRacket can't always determine the proper plot bounds. Here's an example of generating a simple plot of the sine function:

```
(plot (function sin #:color "Blue")
      #:x-min (* -2 pi) #:x-max (* 2 pi)
      #:title "The Sine Function")
```

The result is shown in Figure 4-1. Please note that figures are shown in grayscale, but when you input the code on your computer, graphs will appear in color.

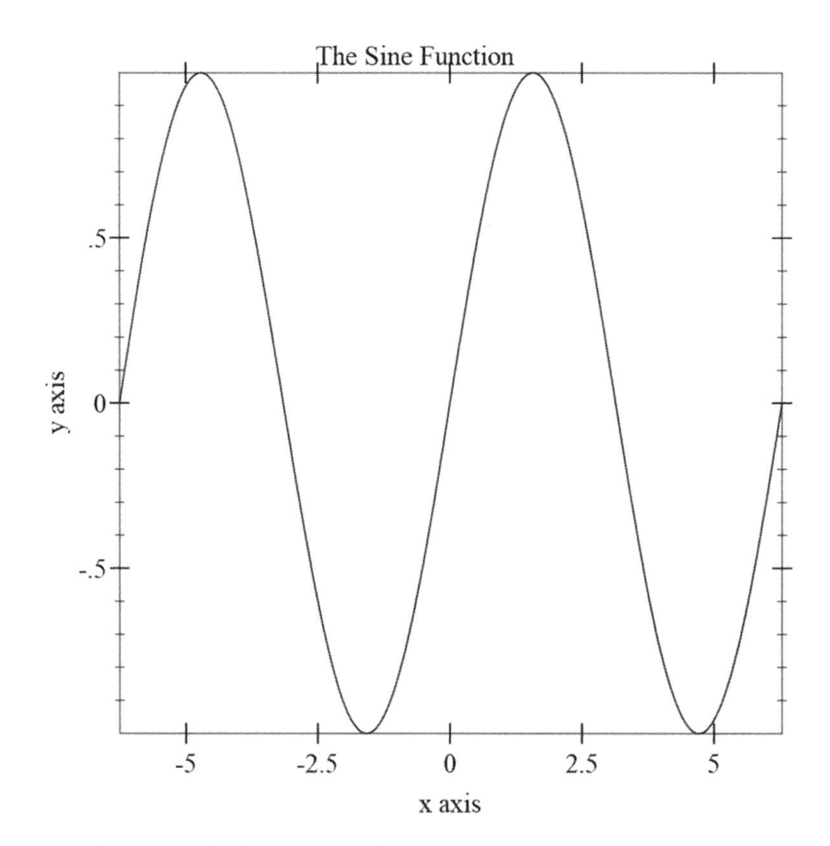

Figure 4-1: Example plot using sine function

Notice that the sin function is enclosed in a function form. The function is something called a *renderer*; it controls how the function argument passed to it gets rendered. The function form allows you to add optional keyword parameters to control how the individual function is displayed. In this case we specify that the sine function is rendered in a blue color.

The following code creates a plot that displays both the sine and cosine functions by combining them into a list.

```
(plot (list
       (axes) ; can also use (axis x y) to specify location
       (function sin #:color "Blue" #:label "sin" #:style 'dot)
       (function cos 0 (* 2 pi) #:color "red"  #:label "cos"))
      #:x-min (* -2 pi) #:x-max (* 2 pi)
      #:y-min -2 #:y-max 2
      #:title "Sine and Cosine"
      #:x-label "X"
      #:y-label #f) ; suppress y-axis label
```

The resulting plot is shown in Figure 4-2. Notice that we've specified a narrower plot range for the cosine function and added text labels to both functions so that they can easily be differentiated on the plot. We've also overridden the default labels with specific values.

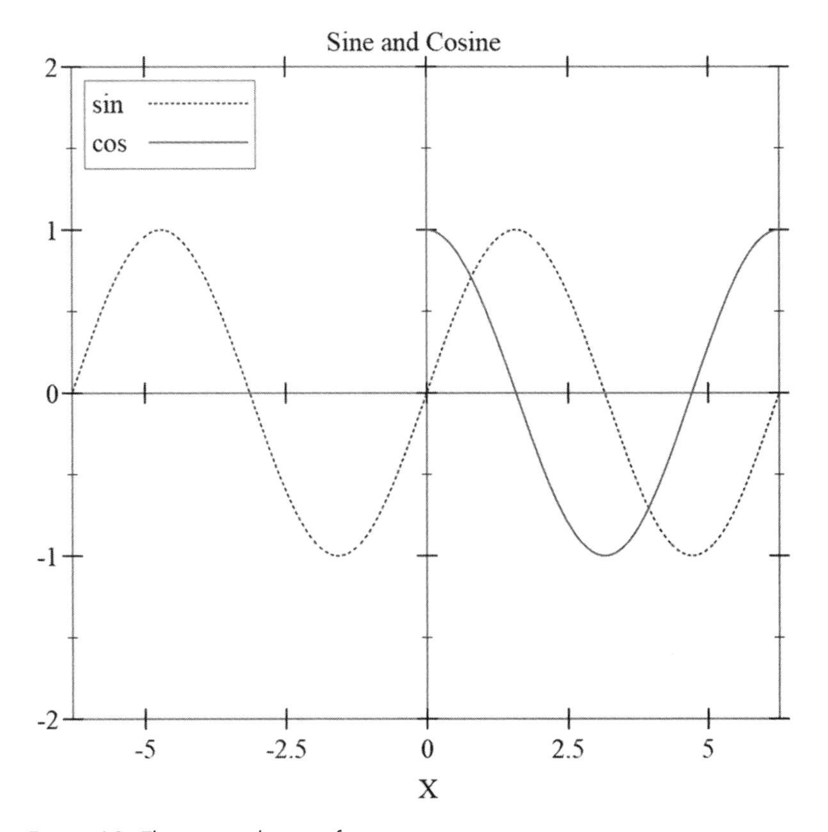

Figure 4-2: The sine and cosine functions

The following code demonstrates some additional plotting capabilities (see Figure 4-3 for the output).

```
(plot (list (axes)
            (function sin #:color "Blue" #:label "sin" #:style 'dot)
            (inverse sqr -2 2 #:color "red" #:label "x^2" #:width 2))
      #:x-min (* -2 pi) #:x-max (* 2 pi)
      #:y-min -2 #:y-max 2
      #:title "Sine and Square"
      #:x-label "X"
      #:y-label #f
      #:legend-anchor 'bottom-left)
```

Notice that we're using a new `inverse` form in our list of functions to plot. This form works by using the y-axis as the axis for the independent variable. In this way, we effectively plot the inverse function without actually having to derive it algebraically. In addition we've specified some additional styles to help differentiate the plot curves. The sine curve is printed dotted, and the inverse square function is printed with a thicker line. The plot legend has been moved to the lower left corner by specifying `#:legend-anchor 'bottom-left` as an optional value to the `plot` function.

If you run the code, you should get something like Figure 4-3.

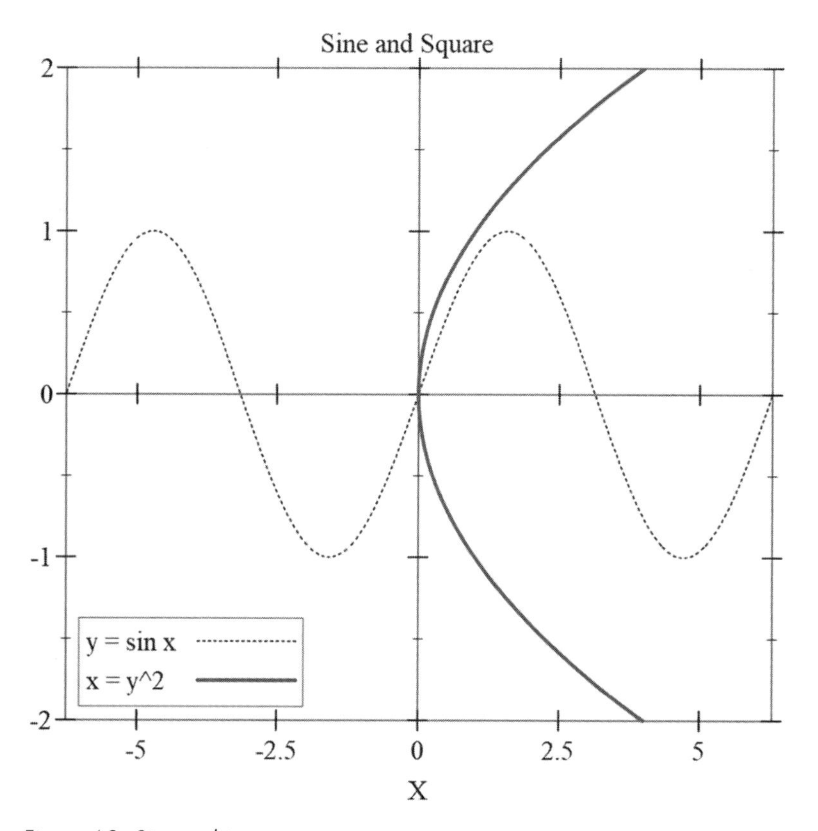

Figure 4-3: Sine and inverse square

Next we take a look at parametric plots.

Parametric Plots

A curve in a plane is said to be parameterized if the coordinates of the curve, (x, y), are represented by functions of a variable (or *parameter*), say t. Let's look at an example.

Parameterizing a Circle

The standard definition of a circle centered at the origin is the set of all points equidistant from the origin. We can state this definition either algebraically as an implicit equation or via a pair of trigonometric expressions as illustrated in Figure 4-4.

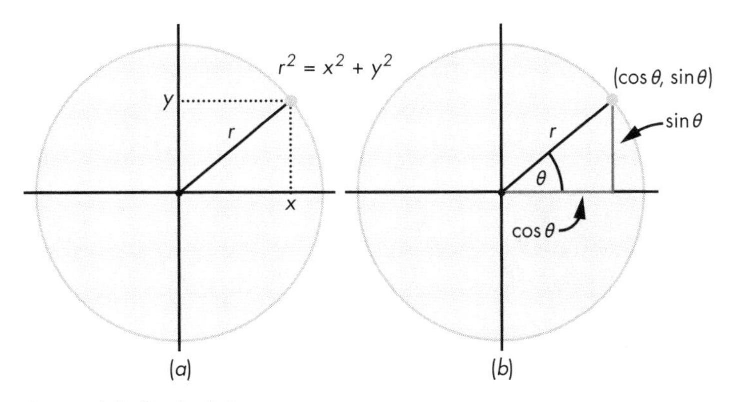

Figure 4-4: Circle definitions

In Figure 4-4 (a), the circle is defined by this algebraic expression:

$$r^2 = x^2 + y^2$$

To plot this as an x-y plot, we have to explicitly express y in terms of x, where the positive values of the expression give the top half of the circle and the negative values give the bottom half:

$$y = \pm\sqrt{r^2 - x^2}$$

Figure 4-4 (b) illustrates defining the circle in terms of the parameter θ. In this case the x- and y-coordinates of the curve are given by the trigonometric expressions $\cos\theta$ and $\sin\theta$ respectively.

In Racket, parametric curves are defined in terms of a function that takes the parameter as its argument and returns a `vector` containing the computed values of x and y. The code below plots both the algebraic and the parametric versions of the circles. We've offset the parametric version so that it appears to the right of the algebraic version (see Chapter 2 for information on the `infix` package and usage of `@` below).

```
#lang at-exp racket
(require infix plot)

(define r 30)
(define off (+ 5 (* 2 r)))

; algebraic half-circles
❶ (define (c1 x) @${ sqrt[r^2 - x^2]})
❷ (define (c2 x) @${-sqrt[r^2 - x^2]})

; parametric circle
❸ (define (cp t) @${vector[off + r*cos[t], r*sin[t]]})

(plot (list
       (axes)
       (function c1 (- r) r #:color "blue" #:label "c1")
       (function c2 (- r) r #:style 'dot #:label "c2")
  ❹ (parametric cp 0 (* 2 pi) #:color "red" #:label "cp" #:width 2))
      #:x-min (- r)
      #:x-max (+ off r)
      #:y-min (- r)
      #:y-max (+ off r)
      #:legend-anchor 'top-right)
```

We've made use of the `infix` package and the `at-exp` language extension so that we can present the algebraic expressions in a more natural form. We begin with definitions of the half-circle functions `c1` ❶ and `c2` ❷. The parametric version is given as `cp` ❸. Notice that `cp` is housed in a form that begins with `parametric` ❹ and a plot range specified as 0 to 2π. The output of this code is shown in Figure 4-5.

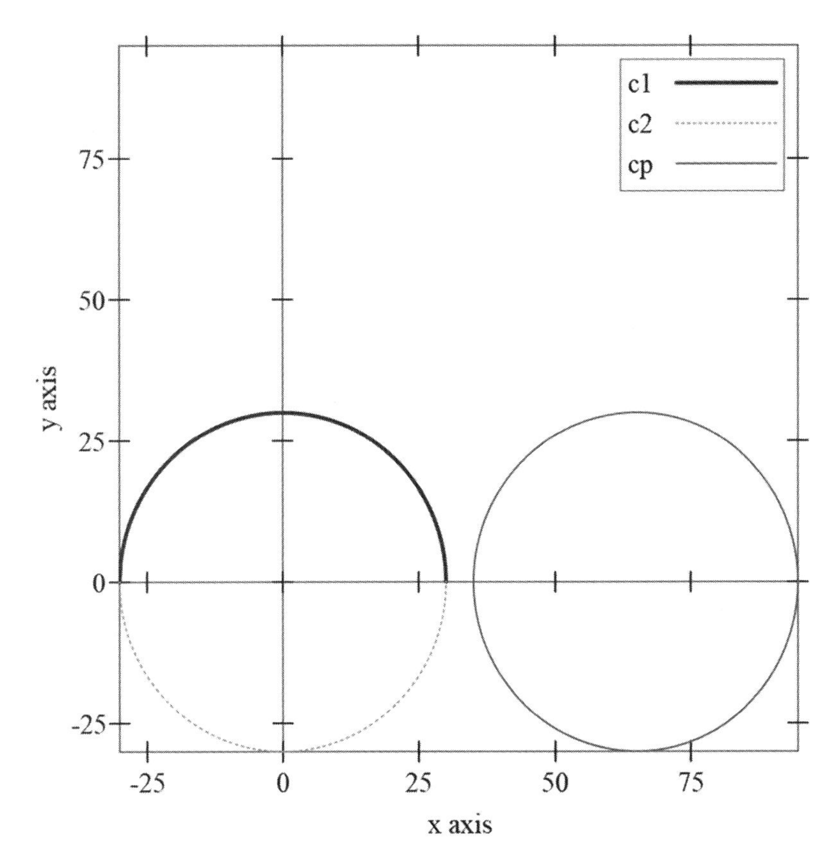

Figure 4-5: Explicit versus parametric plots

Circles in Motion

Let's face it: static circles are pretty boring. In this section we'll take a look at circles in motion. Specifically, we'll examine the path that a fixed point on a circle takes as it rolls along a straight line without slipping. The curve that describes this path is called a *cycloid*. We set the scene in Figure 4-6, where a circle of radius r has rolled distance x from the origin. The point of interest is the dot marked with a P.

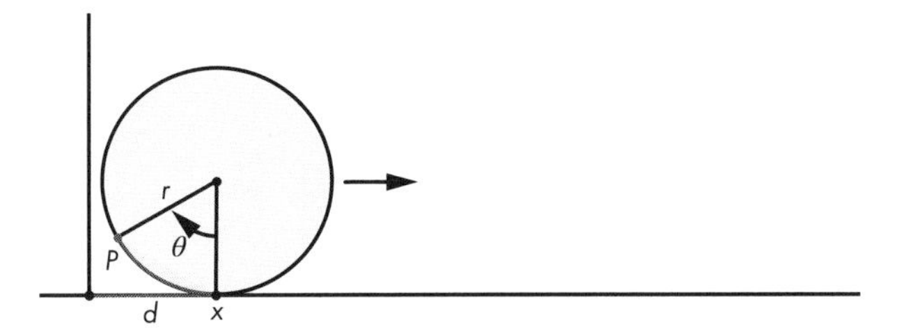

Figure 4-6: Cycloid scene

Let's take a look at the math needed to create a parametric plot of point P that lies on a circle with radius r rolling without slipping on a straight line.

Since the circle rolls without slipping, the length of the arc from x to P is equal to d. We can express this as $d = \theta r$. Thus, as the circle rotates by angle θ, the x-coordinate of P is $d - r\sin\theta$. But $d = r\theta$, so $x = r\theta - r\sin\theta = r(\theta - \sin\theta)$. It's clear that y is simply given by $y = r(1 - \cos\theta)$. Here's the code for the plot:

```
#lang at-exp racket
(require infix plot)

(define r 30)

(define (cycloid t) @${vector[r*(t - sin[t]),  r*(1-cos[t])]})

(plot (list
       (axes)
       (parametric cycloid 0 (* 2 pi)
                   #:color "red"
                   #:samples 1000))
      #:x-min 0
      #:x-max (* r 2 pi)
      #:y-min 0
      #:y-max (* r 2 pi))
```

In this case we've used the `#:samples` keyword to increase the number of samples used to generate the plot, giving it a smoother appearance. Here's the plot.

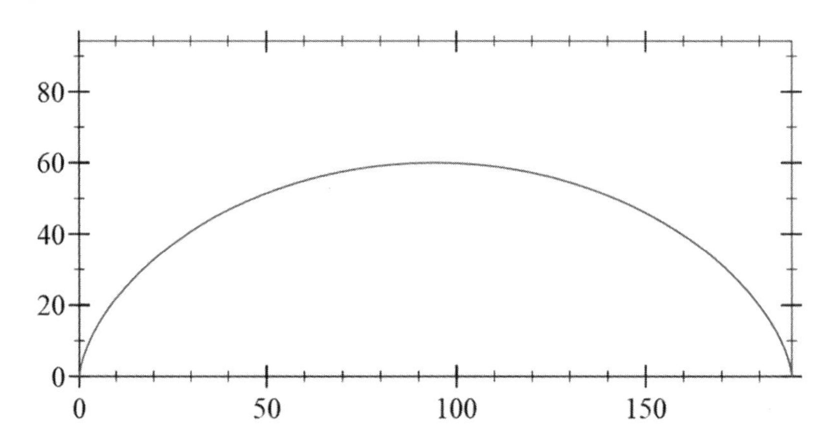

The plot curve has been superimposed on the original scene setup in Figure 4-7.

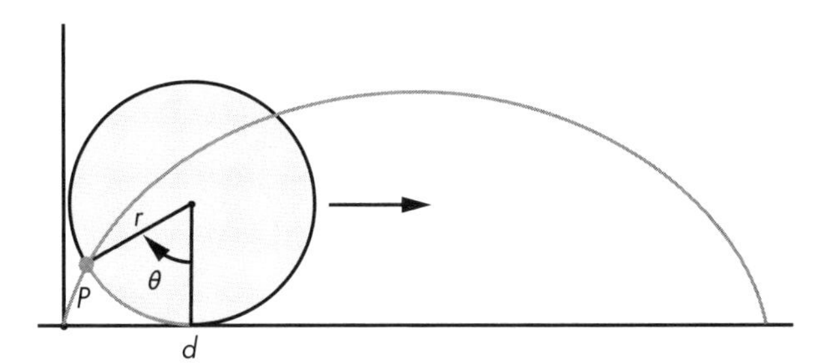

Figure 4-7: Cycloid

Later we'll see how to animate this so that we can actually see it in motion.

The Hypocycloid

Instead of having our circle rolling on a straight line, we could have it rolling on another circle. Indeed, if the one circle is larger than the other, we could have the smaller circle rolling inside the larger circle. A curve generated in this fashion is called a *hypocycloid*. Here we create a parametric plot for a hypocycloid based on a circle with radius r and a larger circle with radius R where the smaller circle rolls without slipping in the interior of the larger circle. We assume that the larger circle is centered at the origin (Figure 4-8).

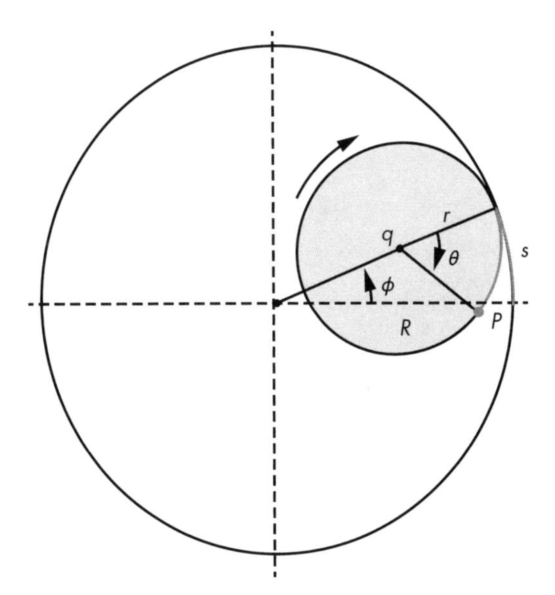

Figure 4-8: Hypocycloid setup

Since the circle rolls without slipping, the arc length s is given by $s = r\theta = R\phi$, so $\theta = \frac{R}{r}\phi$. The center of the smaller circle (q in the diagram) is at a distance $R - r$ from the origin. To assist in our analysis, we've zoomed in on the smaller circle and provided some additional guidelines in Figure 4-9.

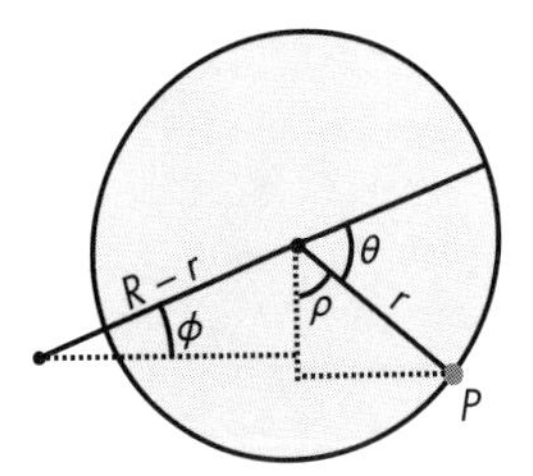

Figure 4-9: Setup zoomed in

Note the following:

$$
\begin{aligned}
\pi &= \left(\frac{\pi}{2} - \phi\right) + \rho + \theta \\
\rho &= \pi - \left(\frac{\pi}{2} - \phi\right) - \theta \\
&= \frac{\pi}{2} + \phi - \theta \\
&= \frac{\pi}{2} + \phi - \frac{R}{r}\phi \\
&= \frac{\pi}{2} + \frac{r - R}{r}\phi
\end{aligned}
$$

From this it's clear that the x-coordinate of P is given by the following:

$$
\begin{aligned}
x &= (R - r)\cos\phi + r\sin\rho \\
&= (R - r)\cos\phi + r\sin\left(\frac{\pi}{2} + \frac{r - R}{r}\phi\right) \\
&= (R - r)\cos\phi + r\cos\left(\frac{r - R}{r}\phi\right) \\
&= (R - r)\cos\phi + r\cos\left(\frac{R - r}{r}\phi\right)
\end{aligned}
$$

Likewise the y-coordinate of P is given by the following:

$$
\begin{aligned}
y &= (R - r)\sin\phi - r\cos\rho \\
&= (R - r)\sin\phi - r\cos\left(\frac{\pi}{2} + \frac{r - R}{r}\phi\right) \\
&= (R - r)\sin\phi + r\sin\left(\frac{r - R}{r}\phi\right) \\
&= (R - r)\sin\phi - r\sin\left(\frac{R - r}{r}\phi\right)
\end{aligned}
$$

We put this all together in the following Racket code:

```
#lang at-exp racket
(require infix plot)

(define r 20)
(define R (* r 3))

(define (hypocycloid phi)
  @${vector[
    (R-r)*cos[phi] + r*cos[(R-r)/r * phi],
    (R-r)*sin[phi] - r*sin[(R-r)/r * phi]]})

(plot (list
       (parametric (λ (t) @${vector[R*cos[t], R*sin[t]]})
                   0 (* r 2 pi)
                   #:color "black"
                   #:width 2)
       (parametric hypocycloid
                   0 (* r 2 pi)
                   #:color "red"
                   #:width 2))
      #:x-min (- -10 R ) #:x-max (+ 10 R )
      #:y-min (- -10 R ) #:y-max (+ 10 R )
      #:x-label #f #:y-label #f
      )
```

You may enjoy the fruits of our labor in Figure 4-10.

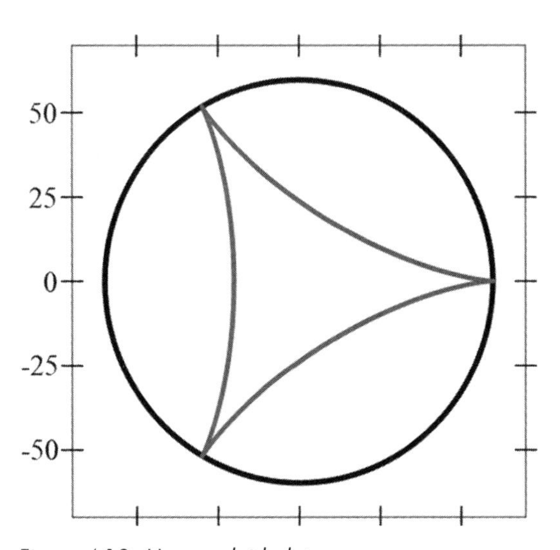

Figure 4-10: Hypocycloid plot

To whet your appetite to further explore these fascinating curves, here's another question for you: for what ratio of radii is a hypocycloid a straight line?

Getting to the Point

So far we've concentrated on generating continuous curves, but what if you want to plot individual points? Yes, Grasshopper, this too is possible. A *point*, from a plotting perspective, is just a vector with two numbers in it. For example, `#(1 2)` is a point. To plot a set of points, you just need to provide the `plot` routine a list consisting of vector points. Here's an example:

```
#lang racket
(require plot)

(parameterize ([plot-width    250]
               [plot-height   250]
               [plot-x-label  #f]
               [plot-y-label  #f])

  (define lim 30)
  (plot (list
         (points '(#(0 0))
                 #:size 300
                 #:sym 'fullcircle1
                 #:color "black"
                 #:fill-color "yellow")
         (points '(#(-5 5) #(5 5))
                 #:size 10
                 #:fill-color "black")
         (points '(#(0 -5))
                 #:size 30
                 #:sym 'fullcircle1
                 #:color "black"
                 #:fill-color "black"))
        #:x-min (- lim) #:x-max (+ lim)
        #:y-min (- lim) #:y-max (+ lim)))
```

And here's the output in Figure 4-11.

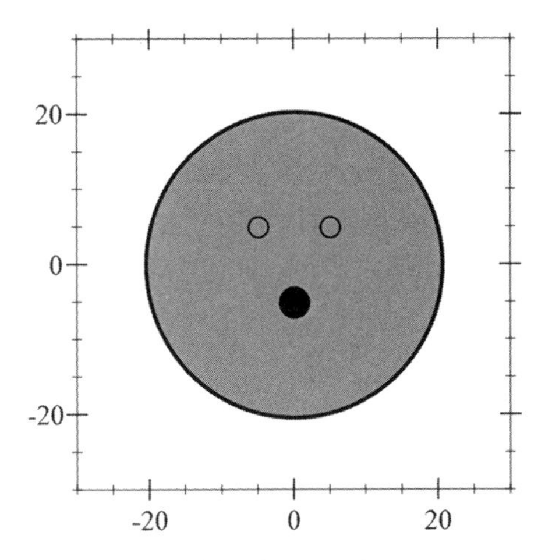

Figure 4-11: A face drawn with points

In this case we're using the parameterize form, which lets us specify the physical size of the plot in pixels (plot-width and plot-height) and suppress the plot labels. Note that the #:size keyword parameter is also expressed in pixels. The first point list is a single point that we use for the face. The #:sym keyword argument specifies what type of symbol to use to print the point. We're using a filled circle specified by the symbol 'fullcircle for the face. There's a large variety of predefined symbols (search for "known-point-symbols" in the documentation), but you may also use a Unicode character or a string.

Now that we have the ability to plot points, we can connect them together with lines using the line form. Like points, lines consist of a list of two-element vectors. The plot routine treats this list as a sequence of line segment end points and plots them appropriately. Here's an example.

```
#lang racket
(require plot)

(define pts (for/list ([i (in-range 0 6)]) (vector i (sqr i))))

(plot (list
       (lines pts
              #:width 2
              #:color "green")
       (points pts
              #:sym 'fulldiamond
              #:color "red"
              #:fill-color "red"))
      #:x-min -0.5 #:x-max 5.5
      #:y-min -0.5 #:y-max 26)
```

The resulting plot is shown in Figure 4-12:

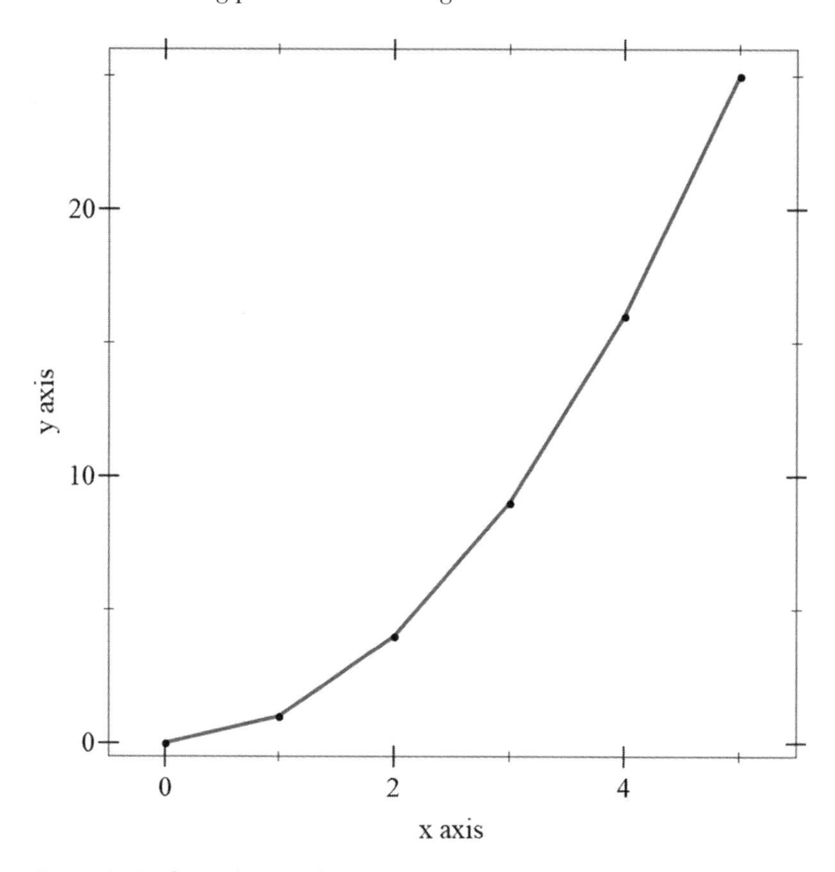

Figure 4-12: Plotting lines and points

Polar Plots

Polar plots are created with . . . (you guessed it!) the `polar` form. The `polar` form is supplied with a function that returns the distance from the origin when supplied with an angle of rotation. The simplest possible function to be used in a polar plot is that of a unit circle, which in the guise of a lambda expression would be `(λ (θ) 1)`. A slightly more ambitious example would be that of a counterclockwise spiral defined as `(λ (θ) θ)`. Here's the code for both plots:

```
(parameterize
    ([plot-width 150]
     [plot-height 150]
     [plot-tick-size 0]
     [plot-font-size 0]
     [plot-x-label  #f]
     [plot-y-label  #f])
```

```
(list (plot (polar (λ (θ) 1) 0 (* 2 pi))
            #:x-min -1 #:x-max 1
            #:y-min -1 #:y-max 1)
      (plot (polar (λ (θ) θ) 0 (* 2.5 pi))
            #:x-min -8 #:x-max 8
            #:y-min -8 #:y-max 8)
      ))
```

We've taken advantage of some plot parameters to suppress the axis tick marks and labels. This code produces the following output in Figure 4-13.

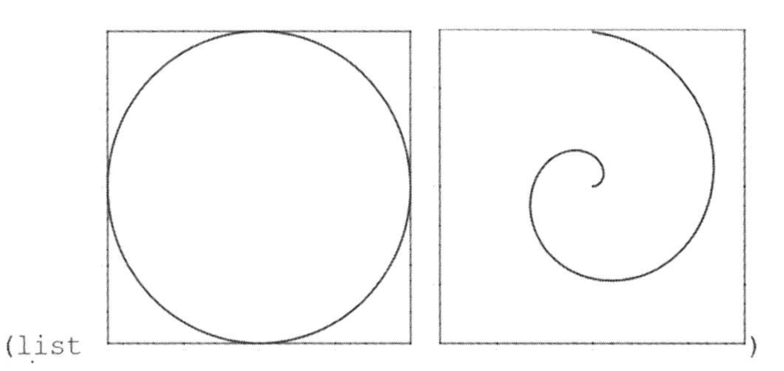

Figure 4-13: Basic polar plots

Notice that the plots are in a list. In Racket a plot is just another data value.

Let's produce the plot shown in Figure 4-14.

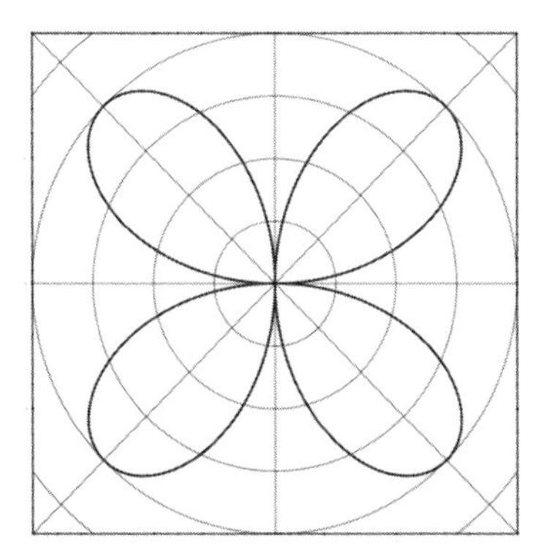

Figure 4-14: The polar rose

We need a function that goes from 0 to 1 and back to 0 again. Sounds like the sine function. We need it to take these values as it goes from 0 to $\pi/2$ (and repeating up to 2π), so the function we need is $r_\theta = \sin(2\theta)$. The code to plot this is then the following:

```
#lang at-exp racket
(require infix plot)

(parameterize
```

```
    ([plot-width 200]
     [plot-height 200]
     [plot-tick-size 0]
     [plot-font-size 0]
     [plot-x-label  #f]
     [plot-y-label  #f])

  (plot (list
        (polar-axes #:number 8)
        (polar (λ (t) @${sin[2*t]})  0 (* 2 pi)
        #:x-min -1 #:x-max 1
        #:y-min -1 #:y-max 1))))
```

With the small addition of parameter k to the function $r_\theta = \sin(k\theta)$, it's possible to produce a large variety of interesting curves (see Figure 4-15).

```
(parameterize
     ([plot-width 200]
     [plot-height 200]
     [plot-tick-size 0]
     [plot-font-size 0]
     [plot-x-label  #f]
     [plot-y-label  #f])

  (define (rose k)
    (plot (polar (λ (t) @${sin[k*t]})  0 (* 4 pi)
                #:x-min -1 #:x-max 1
                #:y-min -1 #:y-max 1)))
  (for/list ([k '(1 1.5 2 2.5 3 4 5)]) (rose k)))
```

Figure 4-15: Polar roses

As another example of a polar plot, here's the code we used to produce the golden spiral first introduced on page 68 in Chapter 3 and reproduced here in Figure 4-16.

```
(define φ (/ (add1 (sqrt 5)) 2))
(define π pi)
(plot (polar (λ (θ) (expt φ (* θ (/ 2 π))))
             0 (* 4 pi)
             #:x-min -20 #:x-max 50
             #:y-min -40 #:y-max 30
             #:color "red")
      #:title "Golden Spiral"
      #:x-label #f ; suppress axis labels
      #:y-label #f)
```

So now we have the actual code used to generate the golden spiral we first encountered in Chapter 3.

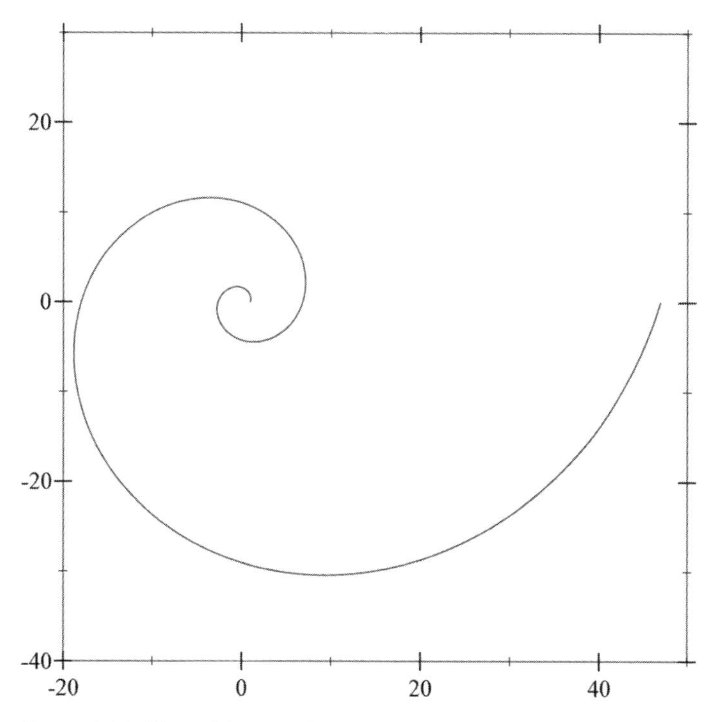

Figure 4-16: The golden spiral

While we've tried to give a good overview of the plotting capabilities that DrRacket affords, there's even more: contours, interval graphs, error bars, vector fields, the list goes on. I encourage you to consult the documentation for additional topics that may be of interest to you. For now, though, we'll be moving on to creating drawings using graphics primitives.

Drawing

Drawing in Racket requires something called a *drawing context (DC)*. You can think of this as a canvas where the drawing takes place. Drawing contexts can be set up for various objects such as bitmaps, PostScript files, or a Racket GUI application.

Unlike coordinates used for plotting functions, drawing contexts use coordinates where the y-axis is inverted as shown in Figure 4-17. Note that the origin is at the upper left corner of the canvas.

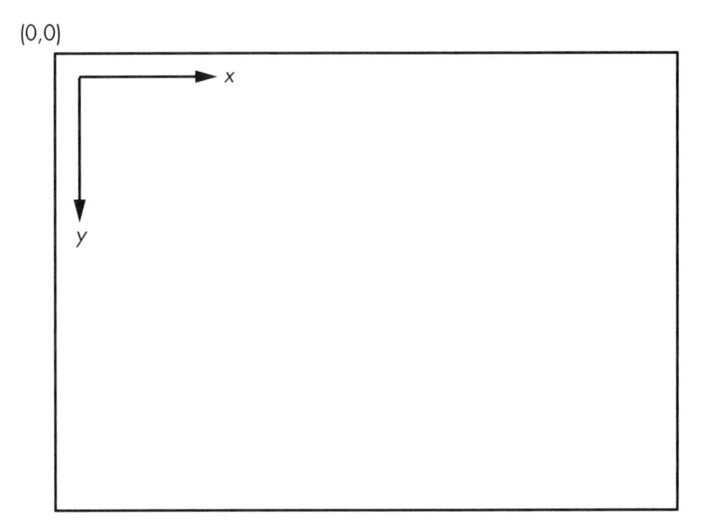

Figure 4-17: Drawing coordinates

All drawings created using the Racket drawing library will need to import the *draw* library, like so:

```
> (require racket/draw)
```

In this section, we'll focus on creating simple drawings using a bitmap drawing context. To create such a context, we need a bitmap (via `make-bitmap` as shown below):

```
> (define drawing (make-bitmap 50 50)) ; a 50x50 bitmap
```

We also need a bitmap drawing context:

```
> (define dc (new bitmap-dc% [bitmap drawing]))
```

Next, we need some drawing tools. In real life we use pens to draw lines and brushes to fill in areas on a canvas; with Racket it's much the same. We can tell Racket the type of pen we want to use in a couple of ways:

```
> (send dc set-pen "black" 2 'solid)
> (send dc set-pen (new pen% [color "black"] [width 2] [style 'solid]))
```

The first method is a quick and dirty way of setting a solid black pen with a width of 2. The second method creates a *pen object* and sends it to the drawing context. Colors can be specified as either a string name as shown above, or a color object where red, green, blue, and optionally alpha (transparency) values can be specified. Each of these values must fall in the range 0–255 (inclusive). For example the following inputs will create a cyan color object and set the drawing context to use it.

```
> (define cyan (make-object color% 0 255 255))
> (send dc set-pen cyan 2 'solid)
```

An equivalent, but slightly more efficient way to perform the above would be the following:

```
> (define cyan (make-color 0 255 255))
> (send dc set-pen cyan 2 'solid)
```

Brushes control the type of fill used inside of two-dimensional objects like rectangles and circles. Like pens, brushes can be defined in a couple of different ways:

```
> (send dc set-brush "blue" 'cross-hatch)
> (send dc set-brush (new brush% [color "red"] [style 'solid]))
```

The first example will create a brush that produces a blue cross-hatch fill. The second example uses a *brush object* that has a solid red fill. An equivalent but slightly different way to achieve the same effect in the second example would be the following:

```
(send dc set-brush (make-brush #:color "red" #:style 'solid))
```

With these preliminaries out of the way, we can actually start drawing by using `send` to send drawing commands to the drawing context. To draw a single line segment, we would input the following:

```
> (send dc draw-line 10 10 30 25)
```

This would draw a line segment beginning at (10, 10) and ending at (30, 25). We have added the line to the drawing context, but it's not immediately displayed. To actually see the line, we do the following:

```
> (print drawing)
```

This produces the following:

Remember that `drawing` was defined earlier and is the actual bitmap we are drawing to.

Rectangles are just as easy to produce:

```
> (send dc draw-rectangle 0 0 50 25)
```

The first two parameters (0, 0) are the x- and y-coordinates of the upper left corner of the rectangle. The next two are the width and height.

Circles and ellipses are handled in a similar fashion.

```
> (send dc draw-ellipse 10 10 30 25)
```

In this case the parameters specify a *bounding box* that contains the ellipse, where again the first two parameters are the x- and y-coordinates of the upper left corner of the bounding rectangle and the next two are its width and height (see the gray area in Figure 4-18—we'll discuss the wedge shape later).

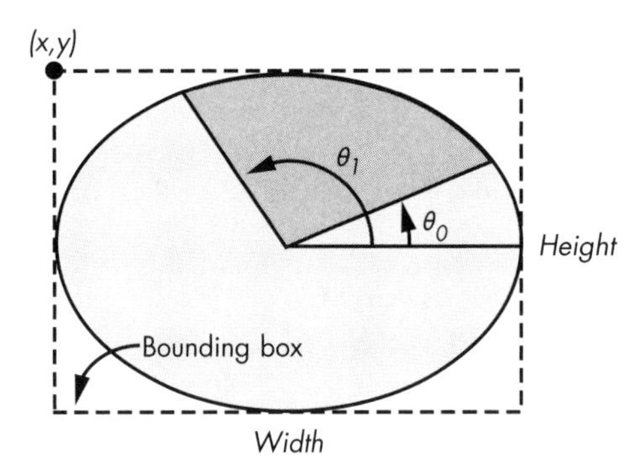

Figure 4-18: Ellipse bounding box

To display the drawing, we simply print the variable (`drawing`) that contains the bitmap we've been drawing to (via the drawing context `dc`):

```
> (print drawing)
```

It produces this:

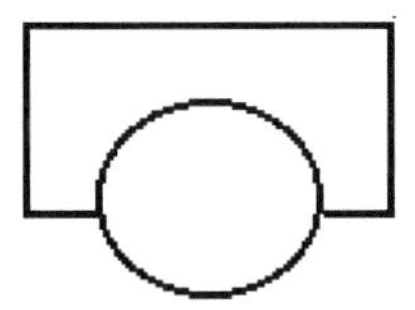

Text can be added to a drawing with the `draw-text` method.

```
> (send dc draw-text "Hello, World!" 10 10)
```

In this case, the last two parameters specify the x- and y-coordinates of the upper left corner of the text.

To pull these ideas together, here is some simple code to draw a crude automobile.

```
#lang racket

(require racket/draw)
```

```
(define drawing (make-bitmap 200 100)) ; a 200x100 bitmap
(define dc (new bitmap-dc% [bitmap drawing]))

; background
(send dc set-brush (new brush% [color "yellow"]))
(send dc draw-rectangle 0 0 200 100)

; antenna
(send dc draw-line 160 5 160 50)
(send dc set-brush (new brush% [color "gray"]))
(send dc draw-rectangle 155 45 10 5)

; body
(send dc set-pen "black" 2 'solid)
(send dc set-brush (new brush% [color "gray"]))
(send dc draw-rectangle 60 20 80 30)

(send dc set-brush (new brush% [color "red"]))
(define c (make-object color% 0 255 255))
(send dc set-pen c 2 'solid)
(send dc draw-rectangle 20 50 160 30)

; wheels
(send dc set-pen "black" 2 'solid)
(send dc set-brush (new brush% [color "blue"]))
(send dc draw-ellipse  40 60 40 40)
(send dc draw-ellipse 120 60 40 40)

(send dc draw-text "This is a car?" 5 1)

(print drawing)
```

Running this code will produce the the stunning work of art shown in Figure 4-19. Depending on your computer the output may look a bit different (primarily due to how fonts are handled).

Figure 4-19: A stunning work of art

The drawing library contains much more functionality than we've illustrated here, but we've learned enough to get started. We'll make practical use of our new skills (and learn some new ones) in the next sections.

Set Theory

You might be thinking that a chapter on graphics is an odd place to host a discussion about set theory, and you'd be right. But it's easier to understand set theory when you can see it, and we can use DrRacket's excellent graphics capabilities to illustrate various basic concepts about set theory. We'll also get to see some additional features of the graphics library.

The Basics

A *set* is just some arbitrary collection of things, such as $\{5, 1/2, 7, 12\}$ or $\{\text{car, bus, train}\}$. One distinguishing feature of a mathematical set is that a mathematical set is not allowed to contain two identical objects. For example $\{8, 2, 9, 2\}$ is *not* a set, since the number 2 occurs twice (such a thing is typically called a *bag* or *multiset*). We can indicate that something is in a set by the *member* (or *in*) symbol: $\in$. For instance $5 \in \{8, 2, 5, 9, 3\}$. Likewise, we can specify that something is not in a set by the not-member symbol: $\notin$, as in $7 \notin \{8, 2, 9, 3\}$.

In Racket, a set can be represented by an object called a *hash set*. A hash set can be either mutable or immutable, depending on how it's constructed. A mutable hash set is created with the `mutable-set` form and an immutable set is constructed with the `set` form. We can test whether an element is a member of a hash set with `set-member?`. Elements can be added to a mutable set by the `set-add!` form and a new set can be created from an old (immutable) one by the `set-add` form.

In the examples that follow, `mset` will designate a mutable set and `iset` will designate an immutable set.

```
> (define mset (mutable-set 5 1/2 7 12 1/2))
> mset
(mutable-set 5 1/2 7 12)

> (set-member? mset 7)
#t

> (set-member? mset 9)
#f

➊ > (set-add mset 9)
. . set-add:
  expected: (not/c set-mutable?)
  given mutable set: (mutable-set 5 1/2 7 12)
  argument position: 1st
```

```
> (set-add! mset 9)
> mset
(mutable-set 5 1/2 7 9 12)

❷ > (set-add! mset 7)
> mset
(mutable-set 5 1/2 7 9 12)

> (define iset (set 3 8 9 7 4))

> iset
(set 9 3 7 4 8)

> (set-add iset 2)
(set 9 3 7 2 4 8)

> (set-add iset 3) ; note, no change in output
(set 9 3 7 2 4 8)

❸ > (set-add! iset 2)
. . set-add!:
  expected: set-mutable?
  given immutable set: (set 9 3 7 4 8)
  argument position: 1st
```

Observe that we were not able to use `set-add` on our mutable set ❶, and we were not able to use `set-add!` on our immutable set ❸. Further, while it did not generate an error, adding the number 7 to `mset` ❷ had no effect since 7 was already a member.

In many mathematical texts, sets are illustrated by using a diagram that contains a rectangle and one or more circles. These diagrams are known as *Venn diagrams*. The rectangle is used to designate all the items of interest (called the *universe of discourse*—we'll use the symbol U to represent this), and circles are used to represent particular sets. To aid in our exploration, we'll define some helper methods to draw various objects in these diagrams.

```
#lang racket
#lang racket
(require racket/draw)

(define WIDTH 150)
(define HEIGHT 100)

(define venn (make-bitmap WIDTH HEIGHT))
(define dc (new bitmap-dc% [bitmap venn]))

(send dc scale 1.0 -1.0)
(send dc translate (/ WIDTH 2) (/ HEIGHT -2))
```

```
(send dc set-smoothing 'smoothed)
(send dc set-pen "black" 2 'solid)

(define IN-BRUSH (new brush% [color "green"]))
(define OUT-BRUSH (new brush% [color (make-object color% 220 220 220)]))
```

The code through (define dc (new bitmap-dc% [bitmap venn])) should be familiar from our previous discussions.

It's a bit inconvenient to have our drawing origin in the upper left corner, with the y-axis inverted, so we use a new feature called a *transformation*. We use the scale transform to scale our drawing environment by 1 in the x-axis direction and -1 in the y-axis direction. This keeps everything the same size, but inverts the y-axis so that up is in the positive direction. To get the origin in the middle of the diagram, we use translate to center it. (There's also a rotate transformation, but it's not needed for our present purposes.)

The set-smoothing argument to send dc enables or disables anti-aliased smoothing for a drawing. The default value of 'unsmoothed produces drawings with a slightly jagged appearance.

The IN-BRUSH will be used as the color to designate things that are in a set, and the OUT-BRUSH is for the color to indicate things that are not in a set.

Next, we'll create a couple of methods to actually do some drawing.

```
(define (rect x y w h b)
  (let ([x (- x (/ w 2))]
        [y (- y (/ h 2))])
    (send dc set-brush b)
    (send dc draw-rectangle x y w h)))

(define (circle x y r b)
  (let ([x (- x r)]
        [y (- y r)])
    (send dc set-brush b)
    (send dc draw-ellipse x y (* 2 r) (* 2 r))))
```

The rect method will draw a rectangle whose center is at coordinates (x, y) with width and height of w and h respectively. And b is the brush that we want to use to draw the rectangle. Similarly, circle will draw a circle whose center is at coordinates (x, y), with radius r, and brush b.

Since we'll only have occasion to draw a single rectangle (representing the universe of discourse, *U*), we create a special function such that it draws it appropriately; we only need to supply it with a brush for the color.

```
(define (universe b) (rect 0 0 (- WIDTH 10) (- HEIGHT 10) b))
```

Let's try these out (see Figure 4-20).

```
> (universe OUT-BRUSH)
> (circle 0 0 30 IN-BRUSH)
> venn
```

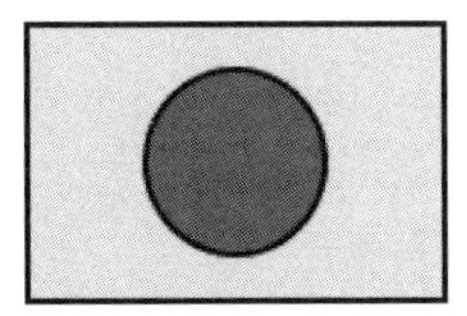

Figure 4-20: A diagram indicating a single set

Suppose our universe of discourse is the integers (that is, $U = \mathbb{Z}$). We can represent the set of even numbers by the green circle. Now suppose that we're interested in anything that is *not* an even number. This is the *complement* of the set of even numbers. The set complement of A can be represented as A^c, $\overline{A}$, or A'. We represent this in a Venn diagram as follows (and shown in Figure 4-21):

```
> (send dc erase)
> (universe IN-BRUSH)
> (circle 0 0 30 OUT-BRUSH)
> venn
```

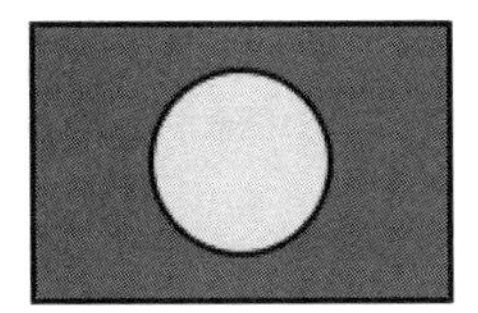

Figure 4-21: Items not in the set

Note that we used `erase` to clear the drawing context before generating the next diagram.

Sets can be combined in various ways. Suppose we have two sets: A and B. One way to combine these sets is to form a new set that is the unique combination of all the elements from set A and B. This operation is called *set union*, and is designated by the symbol $\cup$. If C is the union of A and B, the mathematical expression to reflect this is $C = A \cup B$. Using *set-builder* notation, the set complement would be expressed as $A^c = \{x \in U \mid x \notin A\}$. In words this would be "the set of all x in U such that x is not in A."

Another way to think of the union of sets A and B is to see that it consists of following components:

- All the elements that are in A and not in B (left partial circle).
- All the elements that are in B and not in A (right partial circle).
- All the elements that are in both A and B (the central shape of the diagram—this shape is called a *vesica piscis*, Latin for "bladder of a fish").

See Figure 4-22 for an example.

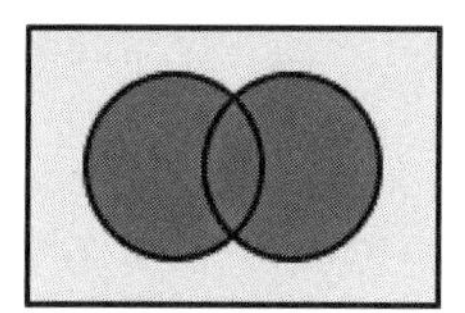

Figure 4-22: Set union

A Short Mathematical Detour

In order to be able to draw the Venn diagram components in Figure 4-22, we'll need to do a few straightforward calculations first. We'll base our calculation on the nomenclature illustrated in Figure 4-23.

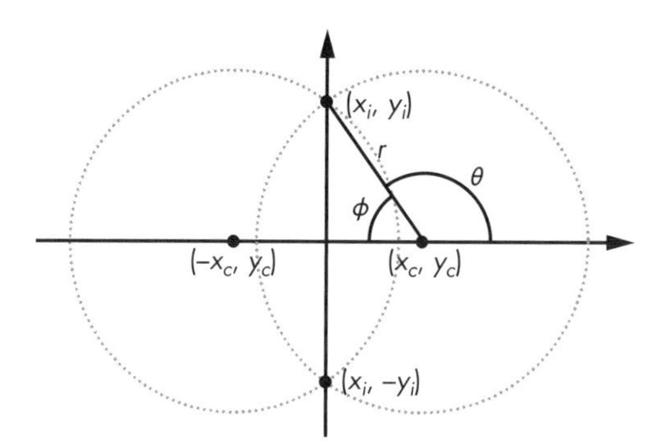

Figure 4-23: How to draw a Venn diagram

Let's see how to find the values of x_i, y_i, θ, and ϕ shown in Figure 4-23. Assume that x_c, y_c, and r are given. Note that the circles are clearly symmetric about the y-axis, which eases our task somewhat. It's immediate that $x_i = 0$. The equation of a circle with center on the x-axis is given by the following:

$$(x - x_c)^2 + y^2 = r^2$$

The intersection points occur at $x = 0$. With this substitution and solving for y, we have the following:

$$y = \pm\sqrt{r^2 - x_c^2}$$

Given y, the angles are easy:

$$\phi = \sin^{-1}\frac{y}{r} \tag{4.1}$$

$$\theta = \pi - \phi \tag{4.2}$$

Now let's exercise a new DrRacket graphics feature called *paths*. The path functionality allows one to draw arbitrary figures. A path is constructed by selecting a starting location and building a sequence of segments to define the entire path. Path segments can consist of straight lines, arcs and something called bezier curves (see the manual). We construct a filled path to represent the center portion of the Venn diagram (the *vesica piscis*) with the following method.

```
(define (piscis x y r b)
  (let* ([y (- y r)]
         [2r (* 2 r)]
         [yi (sqrt (- (sqr r) (sqr x)))] ; y-intersection
         [π pi]
       ❶ [ϕ (asin (/ yi r))]
       ❷ [θ (- π ϕ)]
       ❸ [path (new dc-path%)])
    (send dc set-brush b)
  ❹ (send path move-to 0 (- yi))
  ❺ (send path arc (- x r)     y 2r 2r  θ     (+ π  ϕ))
  ❻ (send path arc (- (- x) r) y 2r 2r (- ϕ) ϕ)
  ❼ (send dc draw-path path)))
```

Within the let* form we find direct translations of Equations (4.1) ❶ and (4.2) ❷. The identifier path is then bound to a new dc-path% object ❸. Paths work sort of like drawing contexts in that path commands are sent to the path object to build the path. The code then positions us at the initial location to draw the first arc ❹ . The first arc is then drawn ❺, and completed by mirroring the first arc ❻. The path arc command works like a drawing context draw-ellipse command. The only difference is that arc takes additional parameters specifying the start and stop angles. The finished path is then sent to the drawing context to be rendered ❼.

Drawing Conclusions

With piscis under our belt, we have most of the machinery we need to draw any binary set operation. To aid in our quest, let's define a simple function to generate the final diagrams.

```
(define SET-BRUSH (new brush% [color (make-object color% 220 255 220)]))

(define (venn-bin b1 b2 b3)
  (universe OUT-BRUSH)
  (circle (- CIRCLE-OFF) 0 30 b1)
  (circle CIRCLE-OFF 0 30 b3)
  (piscis CIRCLE-OFF 0 30 b2)
  (print venn))
```

We define a new light green color in SET-BRUSH to identify the sets participating in the operation. The venn-bin method (the bin part just refers to the fact that it's drawing a binary operation) takes three brushes, one to identify each component of the diagram. The rest of the code should be self-explanatory.

To generate the union diagram we saw in Figure 4-22, we use:

```
> (venn-bin IN-BRUSH IN-BRUSH IN-BRUSH)
```

To illustrate Racket's set operations we'll use two sets:

```
> (define A (set 2 4 6 8 10 12 14 16 18))
> (define B (set 3 6 9 12 15 18))
```

Here's the result of forming the union ($A \cup B$) of the two sets:

```
> (set-union A B)
(set 9 18 14 3 16 2 6 10 15 4 8 12)
```

Notice that sets are not guaranteed to be in any particular order.

Our next operation is *set intersection*. Set intersection is designated by the symbol $\cap$. The intersection of A and B consists of all the elements that are in both A and B. That's $A \cap B = \{x \mid x \in A \text{ and } x \in B\}$. The Venn diagram for intersection is given by the following:

```
(venn-bin SET-BRUSH IN-BRUSH SET-BRUSH)
```

It's also shown in Figure 4-24.

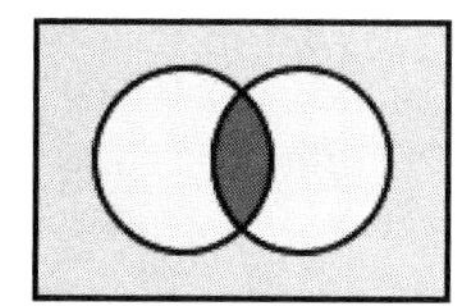

Figure 4-24: Set intersection

Here's an example of Racket code for intersection:

```
> (set-intersect A B)
(set 18 6 12)
```

Next up is *set difference*. Set difference is designated by the symbol \. The set difference of A and B consists of all the elements that are in A that are not also in B. That's $A \setminus B = \{x \mid x \in A \text{ and } x \notin B\}$. The Venn diagram for set difference is given by the following:

```
(venn-bin IN-BRUSH SET-BRUSH SET-BRUSH)
```

It's also shown in Figure 4-25.

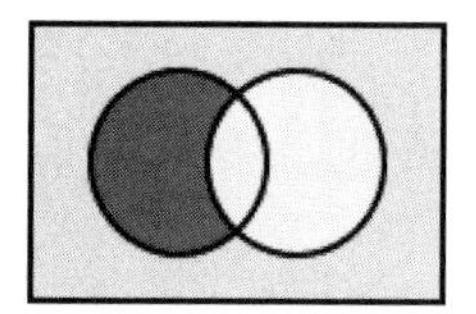

Figure 4-25: Set difference

Set difference is performed with the set-subtract function in Racket.

```
> (set-subtract A B)
(set 14 16 2 10 4 8)
```

Our final operation is *symmetric difference*. Symmetric difference is designated by the symbol $\triangle$. The symmetric difference of A and B consists of all the elements that are in A or B, but are not also in both A and B. That's $A \triangle B = \{x \mid x \in A \text{ or } x \in B, \text{ but not } x \in A \text{ and } x \in B\}$. The Venn diagram for symmetric difference is given by

```
(venn-bin SET-BRUSH IN-BRUSH SET-BRUSH)
```

See an example in Figure 4-26.

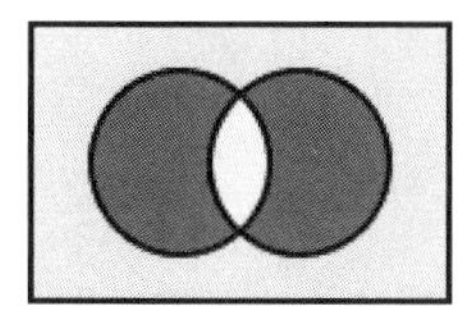

Figure 4-26: Symmetric difference

Symmetric set difference is performed with the set-symmetric-difference function in Racket.

```
> (set-symmetric-difference A B)
(set 9 14 3 16 2 10 15 4 8)
```

Are We Related?

There are a couple of important relationships in the theory of sets that you should be aware of. The first is the concept of *subsets*. A set A is a subset of another set B if all the elements of A are also elements of B. This relationship is represented by the symbol $\subseteq$. In Racket it's possible to test whether one set is a subset of another with the predicate subset?.

```
> (subset? (set 2 4 6 8) (set 1 2 3 4 5 6 7 8 9 10))
#t

> (subset? (set 2 4 6 8) (set 3 4 5 6 7 8 9 10))
#f
```

A subset relationship like the one in the first example can be represented by the Venn diagram shown in Figure 4-27. In this case the inner circle is entirely enclosed by the outer circle.

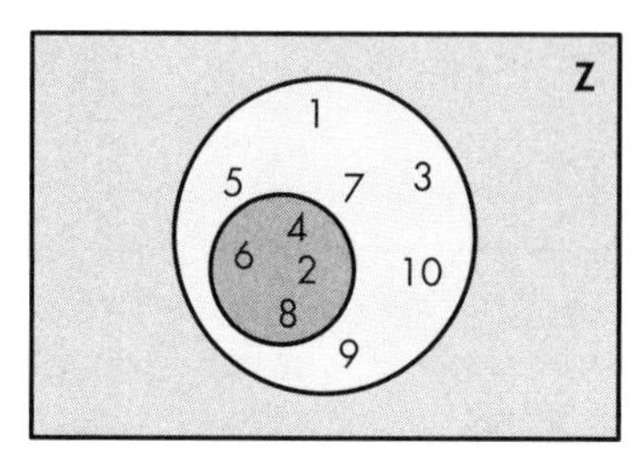

Figure 4-27: A subset relationship

If a set A is a subset of B, but $A \neq B$, A is said to be a *proper subset* of B. This is designated by the symbol $\subset$. Racket provides the `proper-subset?` predicate to perform this function.

```
> (subset? (set 2 4 6 8) (set 2 4 6 8))
#t

> (proper-subset? (set 2 4 6 8) (set 2 4 6 8))
#f
```

Given sets A and B, the *Cartesian product* is defined as the following:

$$A \times B = \{(a, b) \mid a \in A \text{ and } b \in B\}$$

While Racket doesn't have a built-in function that returns the Cartesian product of two sets, it's easy enough to produce our own version.

```
> (define (cart-prod A B)
    (list->set
     (for*/list ([a A]
                 [b B])
       (list a b))))
```

But a somewhat more succinct version is the following:

```
> (define (cart-prod A B)
    (for*/set ([a A]
               [b B])
      (list a b)))
```

which we can test as follows:

```
> (define A (set 'a 'b 'c))
> (define B (set 1 2 3))

> (cart-prod A B)
(set '(a 1) '(c 3) '(c 1) '(c 2) '(a 2) '(a 3) '(b 2) '(b 3) '(b 1))
```

From this it can be seen that the Cartesian product is the set of all possible pairs of values that can be generated from two given sets. Subsets of Cartesian products are frequently used to define relationships. For example, if $A = \{1, 2, 3, 4\}$, we can express the "less than" relationship as follows:

$$\{(1,2),(1,3),(1,4),(2,3),(2,4),(3,4)\}$$

Observe that the first value in each pair is always less than the second element.

This completes our brief foray into applying DrRacket's graphical capabilities to the exciting topic of set theory. Next we'll take a look at a couple of extended applications where we utilize these capabilities to explore additional topics in mathematics.

Applications

Let's begin with an old acquaintance, the Fibonacci sequence.

Fibonacci Revisited

As we learned in the last chapter in the Fibonacci and Friends section, it's always possible to tile a rectangle with squares whose sides are given by the Fibonacci sequence. In this section we'll create a function that can draw the tiling for any Fibonacci number (up to the screen limit). The tiling begins in the center of the drawing canvas and proceeds by essentially spiraling out from that point. Before we dive into writing code for this, a little analysis will guide our way.

Tiling

Observe Figure 4-28. We'll use the upper left corner of each square as a reference point (as indicated by the black quadrants). Each arrow indicates the direction we must move to create the next Fibonacci square.

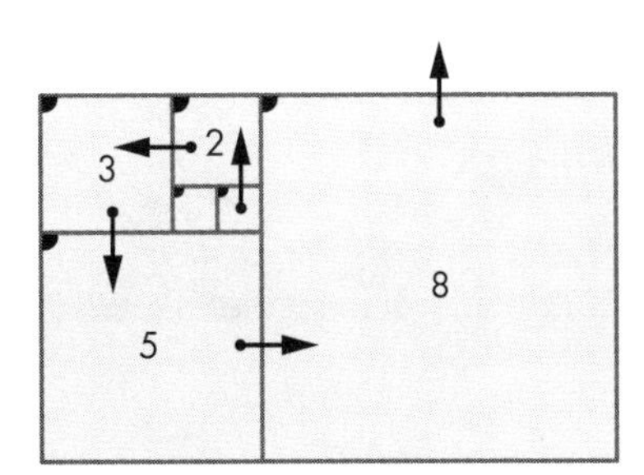

Figure 4-28: Tile analysis

If n is the nth Fibonacci number (F_n), we designate f_n as the nth Fibonacci square with sides of length s_n (this is just some constant multiple of F_n). The coordinates of the upper left corner of f_n will be given by (x_n, y_n). We assume the drawing is initialized with f_1 and f_2 as indicated by the two small, unlabeled squares in the center of the drawing. Moving up from f_2 to f_3, we see that $(x_3, y_3) = (x_1, y_2 - 2)$. We can generalize this to $(x_n, y_n) = (x_{n-2}, y_{n-1} - s_n)$ whenever we move from one tile to another in an upward direction. Table 4-1 gives the change in coordinates for all four directions. Since the pattern repeats itself after four moves, the applicable move is given in the remainder ($n \bmod 4$) column.

Table 4-1: Direction Coordinates

Direction	Coordinates	*n* mod 4
↑	$(x_{n-2}, y_{n-1} - s_n)$	3
←	$(x_{n-1} - s_n, y_{n-1})$	0
↓	$(x_{n-1}, y_{n-1} + s_{n-1})$	1
→	$(x_{n-1} + s_{n-1}, y_{n-2})$	2

This table tells us that at any stage of the drawing process we'll need access to not only F_n and F_{n-1}, but also (x_{n-1}, y_{n-1}) and (x_{n-2}, y_{n-2}). We'll use the second method introduced in the Fibonacci and Friends section to produce F_n in our drawing code. Hence, we'll describe a `draw-tiles` function that takes a Fibonacci n and returns four values: x_{n-1}, y_{n-1}, x_{n-2}, and y_{n-2}. These values along with F_n and F_{n-1} will be used to draw the nth tile.

Figure 4-29 shows the result from calling `(draw-tiles 10)`.

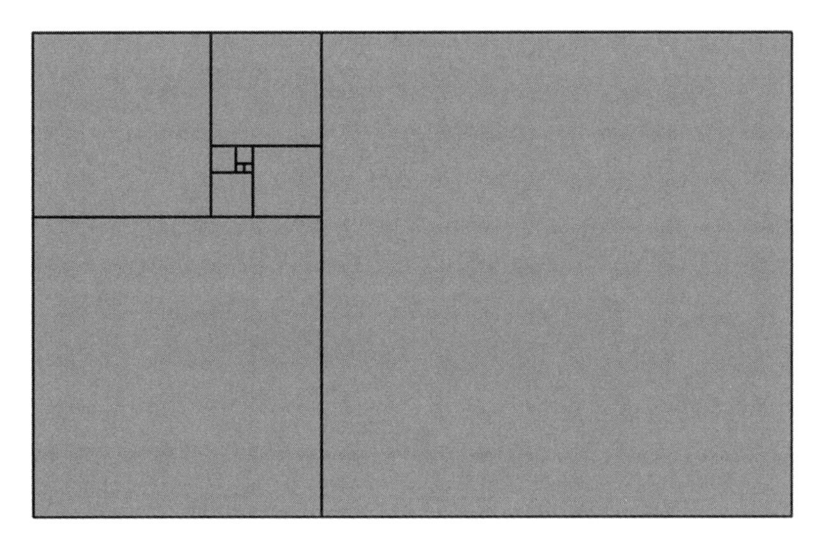

Figure 4-29: A Fibonacci tiling

Our code begins with some constant definitions to establish the parameters to be used in the rest of the program.

```
#lang racket
(require racket/draw)

(define WIDTH 600)  ; width of drawing area
(define HEIGHT 400) ; height of drawing area
(define UNIT 6) ; pixels in unit-width square
(define OFFSET-X 140) ; starting x offset
(define OFFSET-Y 75) ; starting y offset
```

```
(define START-X (- (/ WIDTH 2)  UNIT OFFSET-X))
(define START-Y (- (/ HEIGHT 2) UNIT OFFSET-Y))
```

There should be sufficient comments here to determine what's what. We've fudged a bit by adding some offset values to account for the asymmetric nature of the tiling.

Next we set up our drawing surface, pen, and brush.

```
(define tiling (make-bitmap WIDTH HEIGHT))
(define dc (new bitmap-dc% [bitmap tiling]))

(define TILE-PEN   (new pen% [color "black"] [width 1] [style 'solid]))
(send dc set-pen TILE-PEN)

(define TILE-BRUSH (new brush% [color "yellow"] [style 'solid]))
(send dc set-brush TILE-BRUSH)
```

Now we define two functions: one to compute F_n and the other, `draw-n`, to actually produce the tiling in Listing 4-1:

```
; function to compute F(n)
(define (F n)
  (define (f a b cnt)
    (if (= cnt 0) b
        (f (+ a b) a (- cnt 1))))
  (f 1 0 n))

; function to draw the tiling
(define (draw-n n)
❶ (let* ([fn (F n)]
       ❷ [sn (* UNIT fn)]
       ❸ [fn1 (F (sub1 n))]
       ❹ [sn1 (* UNIT fn1)]
          [n-mod-4 (remainder n 4)])
    (cond [(< n 2) #f] ; do nothing tiles already drawn
          [(= n 2) (values (+ UNIT START-X) START-Y START-X START-Y)]
          [else
        ❺ (let-values ([(x1 y1 x2 y2) (draw-n (sub1 n))])
             (let-values ([(x y)
                           (case n-mod-4
                          ❻ [(0) (values (- x1 sn) y1)]
                             [(1) (values x1 (+ y1 sn1))]
                             [(2) (values (+ x1 sn1) y2)]
                          ❼ [(3) (values x2 (- y1 sn))])])
            ❽ (draw-tile x y sn)
            ❾ (values x y x1 y1)))])))
```

Listing 4-1: Fibonacci with Tiling

The draw-n function is the workhorse of this process. This procedure is recursively called ❺ until the desired number of tiles is drawn. First, we compute F_n ❶ and F_{n-1} ❸. Then we simply multiply these numbers by the constant UNIT ❷ ❹ to determine the sizes of the squares. Next, we determine the coordinates of the upper left corner of the square as described during the discussion of Table 4-1 ❻ ❼. Following this, the tile is actually drawn ❽. Finally, the values x_{n-1}, y_{n-1}, x_{n-2}, and y_{n-2} mentioned in our analysis section above are returned ❾ as these values are needed by the preceding recursive call ❺.

To actually produce the drawing, we have this code:

```
(define (draw-tiles n)
  (draw-tile START-X START-Y UNIT)
  (draw-tile (+ UNIT START-X) START-Y UNIT)
  (draw-n n)
  (print tiling))
```

It initializes the drawing with the two unit squares, calls draw-n, and outputs the constructed bitmap to the screen.

Finally, calling (draw-tiles 10) will produce the output shown previously in Figure 4-29.

The Golden Spiral (Approximation)

To produce the golden spiral discussed in Chapter 3 (and reproduced in Figure 4-30), we only need to make a few additions to our tiling code.

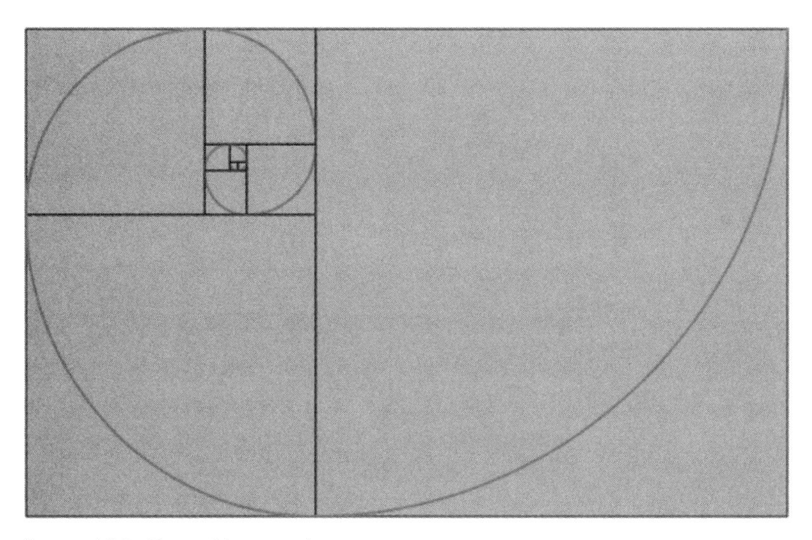

Figure 4-30: The golden spiral

We begin by defining a new drawing pen and brush to be used for creating the spiral.

```
(define SPIRAL-PEN (new pen% [color "red"] [width 2] [style 'solid]))
(define TRANS-BRUSH (new brush% [style 'transparent]))
```

To produce the spiral, we'll be using the draw-arc function. This function works just like draw-ellipse except that it takes two additional arguments that specify the start and stop angles for the arc. These values are designated by θ_0 and θ_1 in Figure 4-18. By default, arcs use a filled brush, so to keep from covering the tiles, we define TRANS-BRUSH as a transparent brush. The spirals are produced by drawing an arc each time a tile is produced. We predefine the various angles (in radians) we will need to use for the arcs:

```
; define angle constants
(define 0d 0)
(define 90d (/ pi 2))
(define 180d pi)
(define 270d (* 3 (/ pi 2)))
(define 360d (* 2 pi))
```

Next we define the actual function used to draw the spiral segments.

```
(define (arc x y r a)
  (let-values ([(d) (values (* 2 r))]
               [(start stop x y)
                (case a
              ❶ [(0) (values  90d 180d x       y   )]
                 [(1) (values 180d 270d x       (- y r)  )]
                 [(2) (values 270d 360d (- x r) (- y r) )]
                 [(3) (values   0d  90d (- x r) y)])])
    (send dc set-pen SPIRAL-PEN)
    (send dc set-brush TRANS-BRUSH)
    (send dc draw-arc x y d d start stop)
    (send dc set-pen TILE-PEN)
    (send dc set-brush TILE-BRUSH)))
```

First, we determine the start and stop angles for the arcs ❶, as well as the x-coordinate and y-coordinate of the location to draw the arcs. Next, we switch to the appropriate pen and brush to draw the arc. Finally, the drawing context is reset to the one needed to draw the tiles.

The only change needed to the draw-n code is to add this single line immediately after the (draw-tile x y sn) statement ❽ in Listing 4-1:

```
(arc x y sn n-mod-4)
```

With this change, calling (draw-tiles 10) will now produce the tiling overlaid by the golden spiral as shown in Figure 4-30.

Nim

Nim is a strategy game in which two players take turns removing objects from distinct heaps. On each turn, a player must remove at least one object, and may remove any number of objects provided they all come from the same heap. The winner of the game is the player who removes the last object. This game can also be played where the person who removes the last object loses, but in our version, the player who takes the last object wins.

In this version of Nim we'll have three piles with up to 15 balls in each pile, as shown in Figure 4-31. Instead of two human players, this time it will be man versus machine, you against the computer, *mano a mano*.

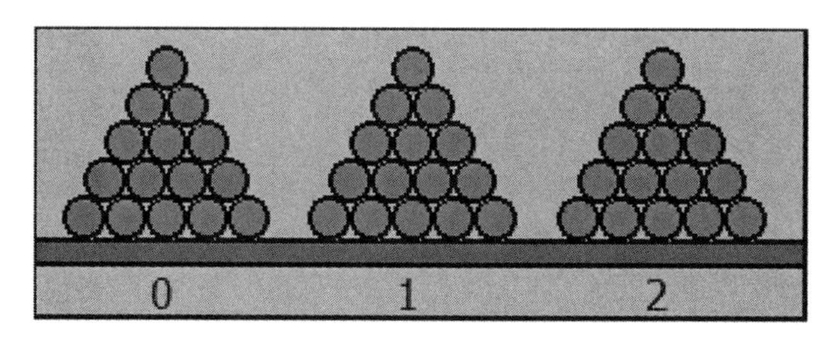

Figure 4-31: Nim starting piles

Setting Up the Graphics

We'll begin with some basic definitions to establish some useful constants. Most of these should be fairly obvious. The `BOARD` references are to the brown board the balls rest on.

```
#lang racket
(require racket/draw)

; overall dimensions of drawing area
(define WIDTH 300)
(define HEIGHT 110)
(define BOTTOM-MARGIN 20)

(define RADIUS 8) ; ball radius
(define DIAMETER (* 2 RADIUS))
(define DELTA-Y (- (* DIAMETER (sin (/ pi 3)))))

(define BOARD-THICKNESS 10)
(define BOARD-Y (- HEIGHT BOARD-THICKNESS BOTTOM-MARGIN))

; location to start drawing pile numbers
(define TEXT-X (+ 5 (* RADIUS 5)))
(define TEXT-Y (- HEIGHT BOTTOM-MARGIN))

; x, y location to start drawing balls
(define START-X 20)
(define START-Y (- BOARD-Y RADIUS))

(define BALL-BRUSH (new brush% [color "red"]))
(define BACKGROUND-BRUSH (new brush% [color "yellow"]))
(define BOARD-BRUSH (new brush% [color "brown"]))
```

The linchpin of Nim's graphics is the draw-pile routine, shown below, which draws a single pile of balls. This code calls draw-ball with the x- and y-coordinates of the location of the center of each ball. It draws each row (see draw-row), bottom up, until there are either no more balls or the row has its full complement of balls.

```
(define (draw-ball x y) ; draw ball with center at (x,y)
  (send dc draw-ellipse (- x RADIUS) (- y RADIUS) DIAMETER DIAMETER))

(define (draw-pile n start-x)
  {let ([rem n]
        [x start-x]
        [y START-Y])
    (define (draw-row x y n max)
      (when (and (> rem 0) (<= n max))
        (set! rem (sub1 rem))
        (draw-ball x y)
        (draw-row (+ x DIAMETER) y (add1 n) max)))
    (for ([r (in-range 5 0 -1)])
      (draw-row x y 1 r)
      (set! x (+ x RADIUS))
      (set! y (+ y DELTA-Y)))})
```

Finally, we come to the code that actually draws the entire Nim environment, draw-game:

```
(define pile (make-vector 3 15))

(define (draw-game)
  (send dc set-pen "black" 2 'solid)
  (send dc set-brush BACKGROUND-BRUSH)
  (send dc draw-rectangle 0 0 WIDTH HEIGHT)
  (send dc set-brush BOARD-BRUSH)
  (send dc draw-rectangle 0 BOARD-Y WIDTH BOARD-THICKNESS)
  (send dc set-brush BALL-BRUSH)

  (draw-pile (vector-ref pile 0) START-X)
  (send dc draw-text "0" TEXT-X TEXT-Y)

  (draw-pile (vector-ref pile 1) (+ START-X (* 6 DIAMETER)))
  (send dc draw-text "1" (+ TEXT-X (* 6 DIAMETER)) TEXT-Y)

  (draw-pile (vector-ref pile 2) (+ START-X (* 12 DIAMETER)))
  (send dc draw-text "2" (+ TEXT-X (* 12 DIAMETER)) TEXT-Y)

  (print drawing) ; display the board
)
(draw-game)
```

This routine simply draws rectangles for the background and board under the piles. It then calls `draw-pile` once for each pile. The number of balls contained in each pile is stored in the vector variable `pile`. For reference, the pile number is shown below each pile.

Gameplay

Nim is a bit unusual in the world of games in that a perfect play strategy is known to exist. It's relatively easy to implement, but it's not obvious. So once you've programmed this, you can spring it on one of your unsuspecting friends and have the computer beat the snot out of them.

The key to this strategy is something called the *nim-sum*. The nim-sum is simply the bitwise exclusive-or of the number of pieces in each pile. The exclusive-or operator is given by the mathematical symbol $\oplus$ and is often designated as xor. It's computed in Racket by the function `bitwise-xor`.

Bitwise exlusive-or is computed as follows: if you are combining two single bits and both bits are the same, then the result is 0; otherwise, it's 1. For example, $1 \oplus 1 = 0$ and $1 \oplus 0 = 1$.

In Racket we can display the binary representation of a number by using `"~b"` in the form string of the `printf` statement. For example, we have the following:

```
> (printf "~b" 13)
1101

> (printf "~b" 9)
1001

> (printf "~b" (bitwise-xor 13 9))
0100
```

Notice that if you take the bitwise exclusive-or in each corresponding bit position of $13_{10} = 1101_2$ and $9_{10} = 1001_2$, you wind up with 0100_2.

It turns out that the winning strategy is just to finish every move with a nim-sum (that is, an exclusive-or) of 0. This is always possible if the nim-sum is not zero before the move. The way to achieve this is with the following:

1. Designate b_0, b_1, and b_2, as the number of balls in piles 0, 1, and 2 respectively.
2. Let $s = b_0 \oplus b_1 \oplus b_2$ be the nim-sum of the all the pile sizes.
3. Compute $n_0 = s \oplus b_0$, $n_1 = s \oplus b_1$, and $n_2 = s \oplus b_2$.
4. Of the three piles, at least one of the values n_0, n_1, or n_2 will be numerically less than the number of items in the corresponding pile. We select from one of these piles and designate it by the letter i.
5. The winning play is then to simply reduce pile i to size n_i. That is, the move is to remove $b_i - n_i$ balls from pile i.

If the nim-sum is zero at the beginning of a player's turn, that player will lose if their opponent plays perfectly. In that case, the best strategy is to stall by picking a single ball from one of the piles and hope that the opponent will make a mistake at some point. The reason this strategy works is somewhat technical, but an analysis can be found in the Wikipedia article at *https://en.wikipedia.org/wiki/Nim*.

This brings us to the actual Racket code to find a winning move.

```
❶ (define nim-sum bitwise-xor)

❷ (define (random-pile) ; select a random pile that has balls
    (let ([i (random 3)])
      (if (> (vector-ref pile i) 0) i (random-pile))))

  (define (find-move)
    (let* ([balls (vector->list pile)]
        ❸ [s (apply nim-sum balls)])
    ❹ (if (= 0 s)
          (let ([i (random-pile)]) (values i 1)) ; stall
        ❺ (let ([n (list->vector (map (λ (b) (nim-sum s b)) balls))])
          ❻ (define (test? i) (< (vector-ref n i) (vector-ref pile i)))
            (define (move i) (values i  (- (vector-ref pile i) (vector-ref n i))
      ))
          ❼ (cond [(test? 0) (move 0)]
            [(test? 1) (move 1)]
            [(test? 2) (move 2)])))))
```

First, we define `nim-sum` as `bitwise-xor` ❶. Next, we have the helper function `random-pile` ❷, which just finds a random pile that has balls. We use this to implement the stalling strategy mentioned above. `find-move` implements our overall playing strategy mentioned in the itemized steps above. This function returns two values: the pile number and the number of balls to remove from the pile. Now we compute the overall nim-sum ❸ (step 2 in the above procedure). This sum is then tested ❹, and if it's zero, it simply returns a random pile with one ball to be removed. The computation mentioned in step 3 above is performed ❺. The local function `test?` ❻ determines whether $n_i < b_i$. The local function `move` returns the pile number and the number of balls to remove as mentioned in step 5 above. Finally, we perform tests to determine which pile to use ❼.

Before we create code for the player to enter their move, we define a few helper functions.

```
(define (apply-move p n) ; remove n balls from pile p
  (vector-set! pile p (- (vector-ref pile p) n)))

(define (game-over?)
  (for/and ([i (in-range 3)]) (= 0 (vector-ref pile i))))

(define (valid-move? p n)
```

```
  (cond [(not (<= 0 p 2)) #f]
        [(< n 0) #f]
        [else (>= (vector-ref pile p) n)]))
```

The apply-move function updates the designated pile by removing the specified number of balls. The game-over? function tests whether there are any balls left to play. And the valid-move? function tests whether a given move is valid.

Tying it all together is the function that implements the following game loop:

```
(define (move p n)
  (if (not (valid-move? p n))
    (printf"\n Invalid move.\n\n")
    (begin (apply-move p n)
           (if (game-over?)
             (printf "\nYou win!")
             (let-values ([(p n) (find-move)])
             (draw-game)
             (printf "\n\nComputer removes ~a balls from pile ~a.\n" n p)
             (apply-move p n)
             (draw-game)
             (when (game-over?)
               (printf "\nComputer wins!")))))))
```

The player enters their move by specifying the pile number followed by the number of balls to remove. For instance, to remove 5 balls from pile 1, one would enter the following:

```
> (move 1 5)
```

To make things interesting, we define an init function that randomly initializes each pile with from 10 to 15 balls.

```
(define (init)
  (for ([i (in-range 3)]) (vector-set! pile i (random 10 16)))
  (newline)
  (draw-game)
  (newline))
```

Figure 4-32 illustrates a game in progress.

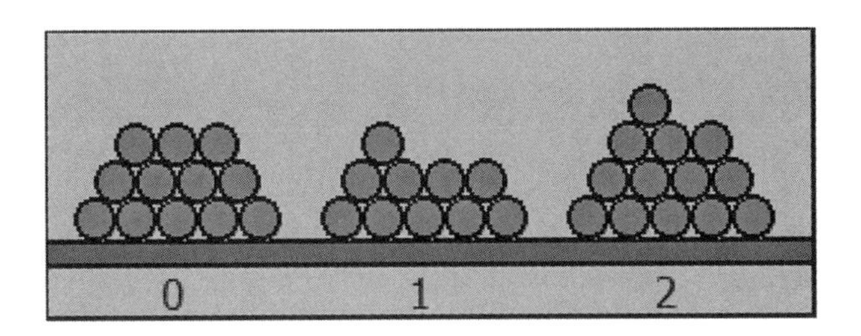


```
> (move 0 1)
```

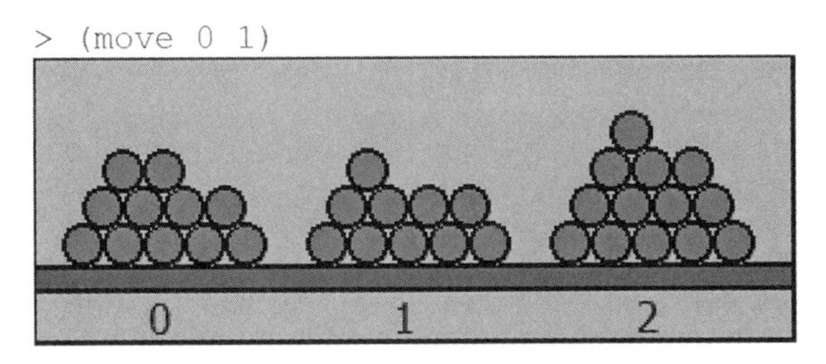


```
Computer removes 4 balls from pile 0.
```

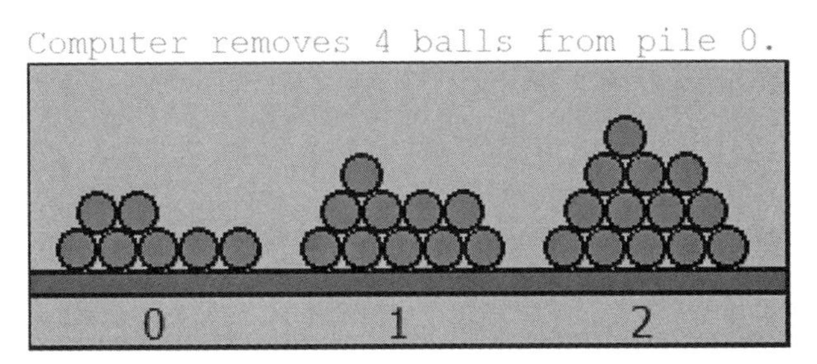

Figure 4-32: Nim: game in progress

Summary

In this chapter, we played with plotting, grappled with graphics, and in the process discovered Racket's extensive capability for visual representation. In the next chapter, we'll build upon this capability to escape the confines of the command line to produce animations and take the first steps toward making interactive applications.

5

GUI: GETTING USERS INTERESTED

GUI (pronounced gooey) stands for graphical user interface. This is just a fancy term for any program that has graphical elements as opposed to just being text-based. Graphical elements can consist of static images or drawings, as we explored in the previous chapter, or interactive elements like buttons and menus. Aside from just graphical elements, GUIs introduce the *event* paradigm—events trigger actions. An event can be anything from a key press or mouse click to a timer going off. Racket not only supports building mini-applications, but also stand-alone executables.

In this chapter, we'll bring our old friend the cycloid to life with an animation, learn some new (card) tricks, and make the Tower of Hanoi more than just an intellectual exercise.

Introduction to GUIs

Racket GUI programming makes use of the *racket/gui/base* library, which can be imported with the (require racket/gui/base) statement. Alternatively, you may use the language switch #lang racket/gui, which includes both the base Racket libraries and the GUI library. Here's a little code snippet that does nothing more than create and display a window frame 300 by 200 pixels.

```
#lang racket/gui

(define frame
  (new frame%
       [label "I'm a GUI!"]
       [width 300]
       [height 200]))

(send frame show #t)
```

You should note that the frame size is the size of the exterior window and includes the size of the title bar and any window borders. The frame interior will be somewhat smaller.

The code below shows something a bit more exciting that demonstrates the basic idea of how to respond to an event.

```
#lang racket/gui

(define main-frame
  (new frame%
       [label "I'm a GUI Too!"]
       [width 300]
       [height 100]))

(define msg
➊ (new message%
     ➋ [parent main-frame]
       [auto-resize #t]
       [label "Hi, there!"]))

➌ (new button%
      [parent main-frame]
      [label "Click Me"]
    ➍ [callback (λ (button event)
                  (send msg set-label "You didn't say may I!"))])

(send main-frame show #t)
```

The message% object ➊ creates a label with the text "Hi, there!" (message% objects are a bit more powerful than you might expect; they can also contain a bitmap as the label). GUI objects must normally specify the parent frame that they are contained in ➋. We've also defined the auto-resize parameter so that the message control will expand if the text is larger than it was initialized with.

Next we create a `button%` object ❸ and make it a child of the `main-frame` window. Buttons can respond to being clicked on with the mouse. This is handled with a `callback` option ❹. This takes a function (in this case a lambda expression) that accepts two parameters. When the button is clicked, the window event processor will pass the callback function a pointer to the button object and an event object that contains information about the event (we don't use them here; we just need to know that the button was clicked). In our case, we send the `msg` object a command for it to set its label to a new value.

Figure 5-1 shows how the thing will look (depending on your operating system, it may be slightly different).

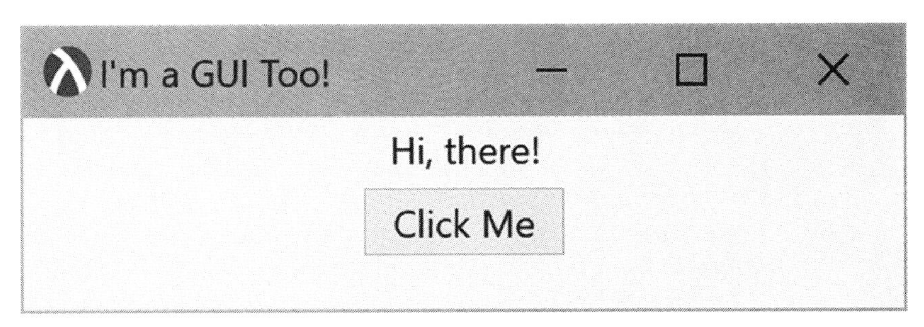

Figure 5-1: Simple GUI application

Let's try our hand at a bit of animation. We'll begin simply, by just moving a little red circle across the screen:

```
#lang racket/gui

(define RADIUS 8)
(define DIAMETER (* 2 RADIUS))

(define loc-x RADIUS)
(define loc-y 35)

(define main-frame
  (new frame%
       [label "I'm a GUI Too!"]
       [width 300]
       [height 100]))

(define canvas
  (new canvas% [parent main-frame]
   ❶ [paint-callback
      (λ (canvas dc)
        (send dc set-smoothing 'smoothed)
        (send dc set-text-foreground "blue")
        (send dc draw-text "Having a ball!" 0 0)
```

```
            (send dc set-brush "red" 'solid)
            (send dc draw-ellipse (- loc-x RADIUS) (- loc-y RADIUS) DIAMETER
        DIAMETER))]))

(define timer
❷ (new timer%
     ❸ [notify-callback
        (λ ()
       ❹ (set! loc-x (add1 loc-x))
       ❺ (send canvas refresh-now))]))

(send main-frame show #t)

❻ (send timer start 50)
```

In this code, we create a canvas% object within the frame% object. This is the same canvas object we explored in the previous chapter, so all the drawing commands we saw previously are available. The canvas object supports a paint-callback event ❶. This event is triggered whenever the canvas needs to be repainted (note that a drawing context, dc, is provided to the callback function). By default it's triggered when the canvas is first displayed, but we force it to be refreshed by using a timer% object ❷. The canvas object draws the ball at (loc-x, loc-y), so our timer will update the loc-x value every time it's called. The timer responds to an event called notify-callback ❸. When this event is triggered, it forces the canvas to refresh itself by sending it a refresh-now message ❺. The timer callback also increments the loc-x variable ❹. This entire process is kicked off with the last line of code. We send the timer object a message to start triggering itself every 50 milliseconds ❻.

We didn't use it here, but the timer can also be stopped by sending it a stop message. The timer% object also supports a just-once? option, if the timer only needs to be triggered a single time. This can also be specified with the start message. For example calling the following code would cause the timer to stop and then to be triggered a single time after waiting one second.

```
> (send timer stop)
> (send timer start 1000 #t)
```

If we wanted the ball to bounce back and forth, we could revise timer as follows:

```
(define timer
  (let ([delta 2])
    (new timer%
         [notify-callback
          (λ ()
            (cond [(<= loc-x RADIUS) (set! delta 2)]
                  [(>= loc-x (- (send canvas get-width) RADIUS)) (set! delta
    -2)])
```

```
        (set! loc-x (+ loc-x delta))
        (send canvas refresh-now))])))
```

This time we define the variable `delta`, which is either positive or negative depending on which direction the ball is moving. We also add a `cond` statement to detect when the ball has reached one of the edges of the canvas and change the direction.

Animating a Cycloid

In the last chapter, we saw how to plot the cycloid curve, which is generated by a point on the perimeter of a circle as it rolls without slipping on a straight line. In this section, we'll produce an animation so that we can see this motion in action. The end result will be an animation of a rolling circle with a point on the circle tracing out the cycloid as it moves along. A snapshot of the animation is shown in Figure 5-2.

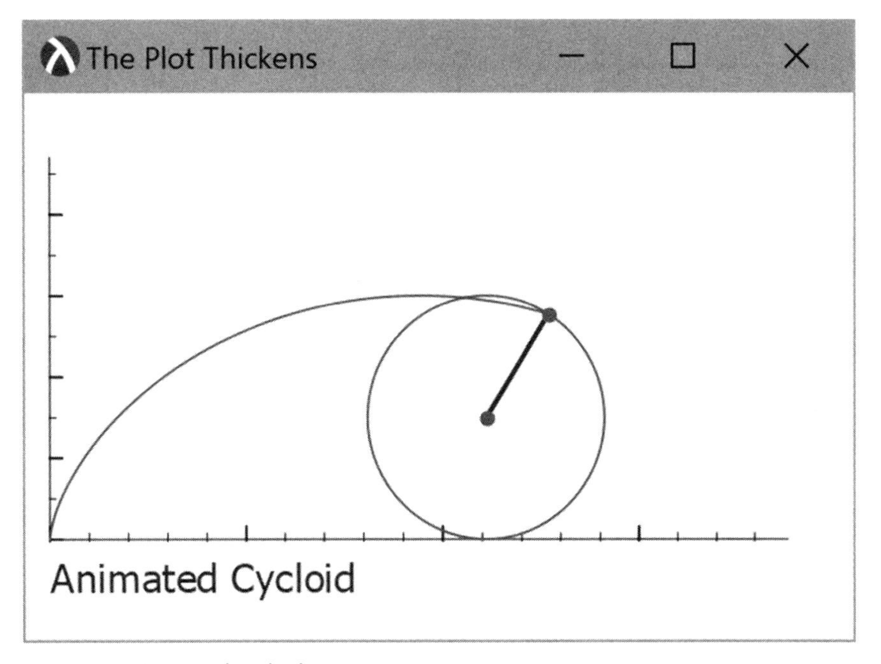

Figure 5-2: Animated cycloid

We once again avail ourselves of the `infix` package to make it easier to enter algebraic expressions. We'll also use the *plot* library that we used before:

```
#lang at-exp racket/gui
(require infix plot)
```

Next we see that the `cycloid` function is unchanged from the one we used before.

```
(define r 30)
(define angle 0)

(define (cycloid t) @${vector[r*(t - sin[t]),  r*(1-cos[t])]})
(define (circle t) @${vector[r*angle + r*sin[t], r + r*cos[t]]})
```

Again r is the radius of the circle used to define the cycloid. The angle variable is used to define the rotation angle at any point in the animation. The circle function will be used to create a parametric plot of the actual circle we're rotating (shown in green in the figure).

To create the blue line segment that runs from the center of the circle to the edge as well as the red end points of the line, we'll use the following two functions. The parameter t is the rotation angle.

```
(define (line t)
  (let ([x @${r*(t - sin[t])}]
        [y @${r*(1 - cos[t])}]
        [x0 (* r angle)]
        [y0 r])
    (lines (list (vector x0 y0) (vector x y))
           #:width 2
           #:color "blue")))

(define (end-points t)
  (let ([x @${r*(t - sin[t])}]
        [y @${r*(1 - cos[t])}]
        [x0 (* r angle)]
        [y0 r])
    (points (list (vector x0 y0) (vector x y))
            #:size 10
            #:sym 'fullcircle1
            #:color "red"
            #:fill-color "red")))
```

There should be no surprises here: *x* and *y* end-point values are computed using basic trigonometry (see the solution to the cycloid problem in the previous chapter). We also examined using lines and points in the previous chapter.

Finally, we get to the actual plot routine:

```
(plot-decorations? #f)

(define (cycloid-plot dc)
  (plot/dc (list
            (axes)
            (parametric circle 0 (* 2 pi) #:color "green")
          ❶ (parametric cycloid 0 angle #:color "red")
          ❷ (line angle)
```

```
❸ (end-points angle))
❹ dc 10 25 300 150
   #:x-min 0
   #:x-max (* r 2 pi)
   #:y-min 0
   #:y-max (* r pi)))
```

In order to plot within the GUI, we must use a special version of `plot` called `plot/dc` since this time we want the output to go to the drawing context. Note the difference between `plot` and `plot/dc` ❹. Here we specify the drawing context, the *x* and *y* location for the plot, and the plot width and height. The other parameters are the same as for `plot`. The bulk of this code specifies the specific objects we want to plot. Of note are: the cycloid itself ❶, the line connecting the circle center to edge ❷, and the line end points ❸. The code `(plot-decorations? #f)`, at the beginning of the listing, turns off some of the axis info so that we have a cleaner plot.

To actually generate the animation, we make some minor changes to the window code we saw earlier:

```
(define main-frame
  (new frame%
       [label "The Plot Thickens"]
       [width 350]
       [height 250]))

(define canvas
  (new canvas% [parent main-frame]
       [paint-callback
        (λ (canvas dc)
          (send dc set-smoothing 'smoothed)
       ❶ (cycloid-plot dc)
          (send dc set-text-foreground "blue")
          (send dc draw-text "Animated Cycloid" 10 180))]))

(define timer
  (new timer%
       [notify-callback
        (λ ()
       ❷ (set! angle (+ 0.1 angle))
       ❸ (when (> angle (* 2 pi)) (set! angle 0))
          (send canvas refresh-now))]))

(send main-frame show #t)

(send timer start 10)
```

Specifially, we update the rotation angle ❷ on each timer tick, and reset the angle to zero ❸ when the circle rotation reaches 2π degrees. The plot actually gets produced with `cycloid-plot dc` ❶.

Having completed a couple of warm-up exercises to familiarize ourselves with some basic GUI functionality, let's take a look at something a bit more ambitious—something we can dazzle our friends with.

Pick a Card

Let's perform a little magic with our computer. In this trick, we deal out a 5-by-5 matrix of cards in some random order, face up. The participant mentally picks one of the cards and indicates which row the card resides in. The cards are then reshuffled and dealt out again in a 5-by-5 matrix. The subject is again asked to pick the row that the card appears on. The selected card is then revealed with great fanfare. Our Racket version of the game is shown in Figure 5-3. The card images used here are available under the GPL[1] thanks to Chris Aguilar.[2]

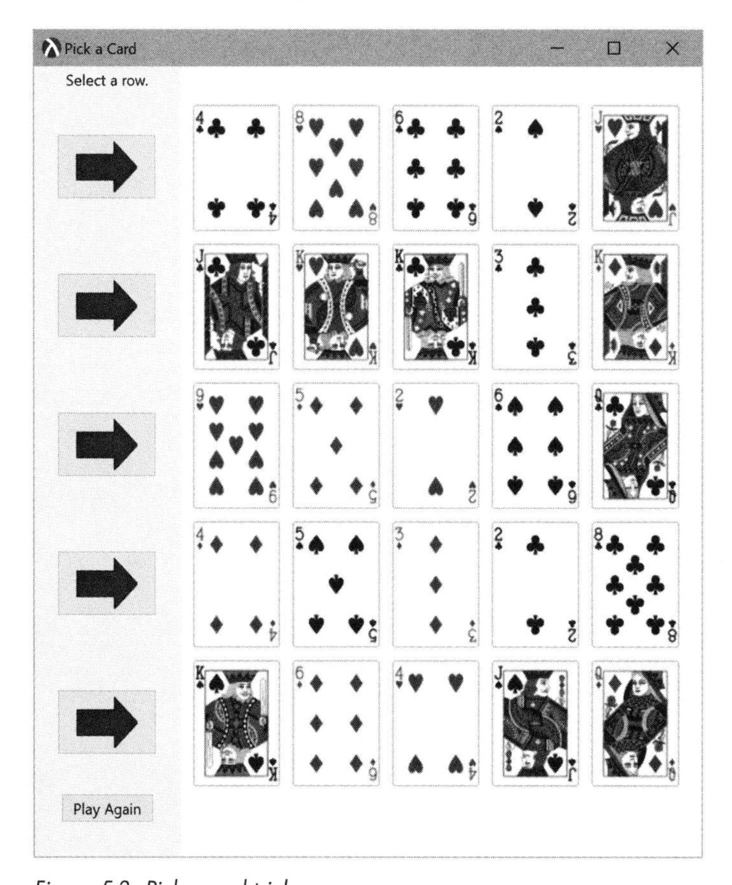

Figure 5-3: Pick-a-card trick

1. GNU General Public License

2. Vectorized Playing Cards 2.0 - *http://sourceforge.net/projects/vector-cards/*. Copyright 2015, Chris Aguilar: *conjurenation@gmail.com*. Licensed under LGPL 3 – *www.gnu.org/copyleft/lesser.html*.

It's clear that if the cards were properly shuffled and re-dealt, it would be impossible to select the proper card without some possibility of error. Like all tricks, this one involves a bit of deception. The mechanism behind this trick is actually quite simple. In the following diagram, we indicate each card by a letter.

	0	1	2	3	4
0	A	B	C	D	E
1	F	G	H	I	J
2	K	L	M	N	O
3	P	Q	R	S	T
4	U	V	W	X	Y

Suppose the `N` is selected in row 2, column 3. We can randomly swap the columns without affecting the selected row. After doing this, we may end up with something that looks like the following.

	0	1	2	3	4
0	C	D	A	E	B
1	H	I	F	J	G
2	M	N	K	O	L
3	R	S	P	T	Q
4	W	X	U	Y	V

We now have something that looks quite different, but we still have the `N` in row 2 (but now in column 1). In an essential step that further adds to the illusion, we transpose the rows and columns (by transpose, we mean the rows become columns and the columns become rows—that is, row 1 becomes column 1, row 2 becomes column 2, and so on). Doing so results in the following arrangement.

	0	1	2	3	4
0	C	H	M	R	W
1	D	I	N	S	X
2	A	F	K	P	U
3	E	J	O	T	Y
4	B	G	L	Q	V

The resulting matrix now looks nothing like the original, but notice that the `N` is now in column 2 instead of row 2. Once the subject indicates row 1 as the proper row (which was originally column 1), we immediately have the location (row 1, column 2—originally row 2). Since we transposed rows and columns, the player is unknowingly revealing the row and column the card resides in.

Swapping rows and columns and transposing rows with columns is standard fare in the mathematical discipline of *linear algebra*. We'll make use of the Racket *matrix* library, which provides the functionality that we need. Given this, our code begins with the following:

```
#lang racket/gui
(require math/matrix)

(define selected-row -1)
(define selected-col -1)
(define show-card #t)
```

The `define` expressions are used to keep track of the program state.

The card bitmaps will be kept in a vector called `card-deck`. The root names of the card images being used are stored in another vector, called `card-names`. The following code reads the card images in from a subfolder called `Card PNGs`:

```
(define card-names
  #("01H" "02H" "03H" "04H" "05H" "06H" "07H" "08H" "09H" "10H" "11H" "12H" "
    13H"
    "01C" "02C" "03C" "04C" "05C" "06C" "07C" "08C" "09C" "10C" "11C" "12C" "
    13C"
    "01D" "02D" "03D" "04D" "05D" "06D" "07D" "08D" "09D" "10D" "11D" "12D" "
    13D"
    "01S" "02S" "03S" "04S" "05S" "06S" "07S" "08S" "09S" "10S" "11S" "12S" "
    13S"))


(define card-deck
  (for/vector ([card-name (in-vector card-names)])
```

```
    (read-bitmap (build-path "Card PNGs" (string-append card-name ".png")))))

(define card-width (send (vector-ref card-deck 0) get-width))
(define card-height (send (vector-ref card-deck 0) get-height))
```

Note that we've used build-path to construct an operating system–agnostic pathname.

Since we'll want to be able to shuffle the cards, the following code will place the card-deck in a random order:

```
(define (shuffle-deck)
  (for ([i (in-range 52)])
    (let ([j (random 52)]
          [t (vector-ref card-deck i)])
      (vector-set! card-deck i (vector-ref card-deck j))
      (vector-set! card-deck j t))))
```

The card-deck vector is just a linear list of the card bitmaps. To be able to present them in the arrangement to be shown on the display, we define a Racket display-matrix that holds the indexes into the card-deck vector. We also create get-card, which enables us to access any bitmap given the row and column it resides in.

```
(define display-matrix
  (build-matrix SIZE SIZE (λ (r c) (+ (* r SIZE) c))))

(define (get-card r c)
  (vector-ref card-deck (matrix-ref display-matrix r c)))
```

In this section, we've explained the basic mechanism behind the trick and defined some data structures to hold the puzzle objects (the cards and their images). In the next section, we'll exploit Racket's layout mechanism to generate an attractive tableau.

GUI Layout

We've used buttons, the message control, and the canvas control before, but we displayed them in the default order, which just stacks one on top of the other. In this case we need something a bit more sophisticated. Racket provides two layout controls that are sufficient for our purposes: horizontal-panel% and vertical-panel%. The vertical-panel% control allows us to stack controls vertically, as shown on the left side of Figure 5-3. In the code below, we use the horizontal-panel% to house the vertical-panel% and the canvas used to draw the card faces.

```
(define main-frame
  (new frame%
       [label "Pick a Card"]
       [width 550]
       [height 650]))
```

```
(define main-panel (new horizontal-panel%
                        [parent main-frame]))

(define control-panel (new vertical-panel%
                           [parent main-panel]
                           [min-width 100]
                           [stretchable-width 100]))

(define MARGIN 10)     ; in pixels
(define SIZE 5)        ; card rows and columns
(define MSG-HEIGHT 20) ; height of msg label

(define canvas
  (new canvas%
       [parent main-panel]
       [min-width 400]
       [paint-callback
        (λ (canvas dc)
          (send dc set-smoothing 'smoothed)
          (for* ([r (in-range SIZE)] ; draw the cards
                 [c (in-range SIZE)])
            (send dc draw-bitmap (get-card r c)
                  (+ MARGIN (* c (+ MARGIN card-width)))
                  (+ MSG-HEIGHT MARGIN (* r (+ MARGIN card-height)))))
       ❶ (when show-card ; draw red border on selected card
            (let* ([off-x (/ MARGIN 2)]
                   [off-y (+ off-x MSG-HEIGHT)])
              (send dc set-pen "red" 3 'solid)
              (send dc set-brush (new brush% [style 'transparent]))
              (send dc draw-rectangle
                    (+ off-x (* selected-col (+ MARGIN card-width)))
                    (+ off-y (* selected-row (+ MARGIN card-height)))
                    (+ card-width MARGIN) (+ card-height MARGIN))
              (send dc set-pen "black" 2 'solid)))
          )]))
```

Most of the code for canvas should be familiar, but the section starting with when ❶ uses a state variable called show-card that's set to true when it's time to reveal the selected card (via a red border around the card).

Building the Controls

We want a message control to be shown in the upper left portion of the application to act as a prompt to the user. Let's add the code to do that now. Begin with the following code.

```
(define msg
  (new message%
       [parent control-panel]
       [min-height MSG-HEIGHT]
       [label "Select again."]))
```

To dress up our buttons, we need an image of an arrow. Rather than go out and try to find a bitmap of an arrow, it's a simple matter to construct one on the fly, as we do here via a sequence of `path` statements to generate a polygon in the shape of an arrow:

```
(define arrow ; bitmap
  (let* ([image (make-bitmap 50 40)]
         [dc (new bitmap-dc% [bitmap image])]
         [path (new dc-path%)])
    (send dc set-brush (new brush% [color "blue"]))
    (send path move-to  0 10)
    (send path line-to 30 10)
    (send path line-to 30  0)
    (send path line-to 50 20)
    (send path line-to 30 40)
    (send path line-to 30 30)
    (send path line-to  0 30)
    (send path line-to  0 10)
    (send dc draw-path path)
    image))
```

Here's a slightly more concise way to write this:

```
(define arrow ; bitmap
  (let* ([image (make-bitmap 50 40)]
         [dc (new bitmap-dc% [bitmap image])]
         [path (new dc-path%)])
    (send dc set-brush (new brush% [color "blue"]))
    (send path move-to  0 10)
    (send path
          lines '(
                  (30 . 10)
                  (30 . 0)
                  (50 . 20)
                  (30 . 40)
                  (30 . 30)
                  ( 0 . 30)
                  ( 0 . 10)))
    (send dc draw-path path)
    image))
```

Now that we have our arrow, we also need buttons to select the card row. We'll use the following code:

```
(define (gen-row-button r)
  (new button%
       [parent control-panel]
       [label arrow]
       [min-width 80]
       [min-height 50]
       [vert-margin (/ (+ MARGIN (- card-height 50)) 2)]
       [callback (λ (button event)
                   (select-row r))]))

(for ([i (in-range SIZE)])
  (gen-row-button i))
```

Note we've used the card height to adjust the button margins so that they line up properly with the row of card images.

Control Logic

Now that we've specified the basic GUI components, we move on to the control logic where we make the puzzle interactive. The logic to swap the columns of the `display-matrix` is given in the following `swap-cols` function. It takes two column numbers and swaps the columns of `display-matrix` as required. We're going to say a few words about how this works, but if you're not familiar with linear algebra, you may want to take this function as a black box that works as expected and skim over the next section.

```
(define (swap-cols c1 c2)
  (let ([swap-matrix (make-swap c1 c2)])
    (matrix* display-matrix swap-matrix)))
```

Linear Algebra Zone

As you will recall from linear algebra, there are three elementary matrix operations.

- Interchange two rows (or columns).
- Multiply each element in a row (or column) by a number.
- Multiply a row (or column) by a number and add the result to another row (or column).

All these operations can be performed by matrix multiplication. For our purposes we only need to implement the ability to interchange columns, but if M and R are matrices such that $RM = M_r$, where certain rows of M are swapped, then $MR = M_c$ produces a matrix with the corresponding columns swapped. As an example we define M and R as follows:

$$M = \begin{bmatrix} A & B & C \\ D & E & F \\ G & H & I \end{bmatrix}$$

$$R = \begin{bmatrix} 0 & 1 & 0 \\ 1 & 0 & 0 \\ 0 & 0 & 1 \end{bmatrix} \tag{5.1}$$

Then we can swap the first two rows or first two columns of M as follows:

$$RM = \begin{bmatrix} 0 & 1 & 0 \\ 1 & 0 & 0 \\ 0 & 0 & 1 \end{bmatrix} \begin{bmatrix} A & B & C \\ D & E & F \\ G & H & I \end{bmatrix} = \begin{bmatrix} D & E & F \\ A & B & C \\ G & H & I \end{bmatrix}$$

$$MR = \begin{bmatrix} A & B & C \\ D & E & F \\ G & H & I \end{bmatrix} \begin{bmatrix} 0 & 1 & 0 \\ 1 & 0 & 0 \\ 0 & 0 & 1 \end{bmatrix} = \begin{bmatrix} B & A & F \\ E & D & C \\ H & G & I \end{bmatrix}$$

Racket supplies a matrix multiplication operator, `matrix*`, but unfortunately does not provide built-in row or column swapping operations, so we must create our own. The following `make-swap` function takes two row numbers and returns a matrix with the corresponding matrix rows swapped. We employ this in conjunction with the Racket-supplied matrix multiplication operator in the `swap-cols` function given above. Since `swap-matrix` is the second argument in the multiplication, it performs a column swap instead of a row swap.

```
(define (make-swap r1 r2)
  (define (swap-func r c)
    (cond [(= r r1) (if (= c r2) 1 0)]
          [(= r r2) (if (= c r1) 1 0)]
          [(= r c) 1]
          [else 0]))
  (build-matrix SIZE SIZE swap-func))
```

The function `build-matrix` is defined in the Racket *matrix* library. It constructs a new matrix by populating it with elements computed by `swap-func`.

Wrapping Up the GUI

When the user makes their first row selection, the following code runs:

```
(define (first-row-selection r)
  (set! selected-col r)
  (send msg set-label "Select again.")
❶ (for ([i (in-range SIZE)])
    (let ([j (random SIZE)]
          [t (vector-ref card-deck i)])
   ❷ (set! display-matrix (swap-cols i j))))
❸ (set! display-matrix (matrix-transpose display-matrix))
  (send canvas refresh-now))
```

You may be wondering why the first thing it does is set `selected-col` with the selected row `r`. The reason is that the selected row will become the selected column after the built-in `matrix-transpose` is executed ❸. We perform a series of column swaps to jumble up the displayed equation without affecting the row order ❶ ❷).

The second time the user makes a row selection, the `show-selection` function runs.

```
(define (show-selection r)
  (send msg set-label "Tada!")
  {set! selected-row r}
  (set! show-card #t)
  (send canvas refresh-now))
```

This function sets the `show-card` variable and triggers the canvas to refresh and reveal the selected card.

To initialize the entire process, we have the `restart` function:

```
(define (restart)
  (shuffle-deck)
  (send msg set-label "Select a row.")
  (set! show-card #f)
  (set! selected-row -1)
  (set! selected-col -1)
  (send canvas refresh-now))

(restart)

(send main-frame show #t)
```

Let's see what happens when a user clicks on a button. Way back when we created the arrow buttons, we assigned them a callback function called `select-row` along with the appropriate row number. Depending on the state variables `selected-row` and `selected-col`, the function will perform different actions.

```
(define (select-row r)
  (cond [(< selected-col 0) (first-row-selection r)]
        [(< selected-row 0) (show-selection r)]
        [else (send msg set-label "Restart.")]))
```

If `selected-col` is less than zero (meaning this is the first selection), it runs `first-row-selection`. If `selected-row` is less than zero (meaning this is the second selection), it runs `show-selection` to reveal the selected card. If neither of these are true (meaning both selections have been made), it prompts the user to reset the program by pressing the Play Again button.

Our card trick application has exercised much of Racket's GUI capabilities, but there's nothing quite like seeing objects flying around the screen. In the next section, we'll convert our command line version of the Tower of Hanoi into an interactive animated puzzle.

Control Tower

Back in Chapter 3, we looked at a simple function that solved the Tower of Hanoi puzzle. The function returned a list that contained the moves required to solve the puzzle. This is of course not a very satisfying way to visualize how the solution progresses. In this section we'll remedy that omission by animating the solution and learn a bit more about Racket widgets in the process. Widgets are graphical objects used to provide input to an application (such as buttons, text boxes, and pick lists) or display information (such as progress bars and labels).

The code to solve the puzzle (`hanoi`) is repeated below. It still does exactly what we need. Recall that it returns a list of moves, each of which is a list containing the peg to move from and the peg to move to. It can be placed anywhere after the `#lang racket/gui` command.

```
(define (hanoi n f t)
  (if (= 1 n) (list (list f t))
      (let* ([u (- 3 (+ f t))] ; determine unused peg
             [m1 (hanoi (sub1 n) f u)] ; move n-1 disks from f to u
             [m2 (list f t)] ; move single disk from f to t
             [m3 (hanoi (sub1 n) u t)]); move disks from u to t
        (append m1 (cons m2 m3)))))
```

The program we'll build in the following sections will accommodate from 1 to 10 disks. Figure 5-4 illustrates some basic parameters we'll be using in the rest of this section. As shown, each peg will be assigned a number from 0 to 2. Each position on a peg also has a designated value. These values will be used as parameters to various functions to control the source and target location of the disks.

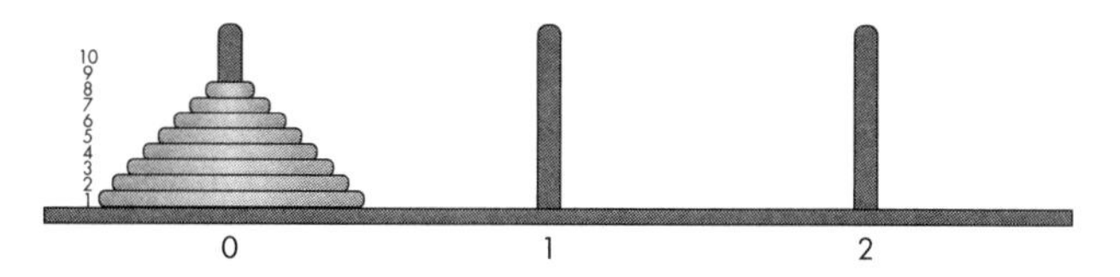

Figure 5-4: Tower of Hanoi parameters

Setting Up

We'll begin with some useful constants.

```
#lang racket/gui

(define MAX-DISKS 9)
(define UNIT 15)
(define PEG-X (+ (* 3 UNIT) (* MAX-DISKS (/ UNIT 2))))
(define PEG-Y (* 2 UNIT))
(define START-Y (+ PEG-Y (* UNIT MAX-DISKS)))
(define PEG-DIST (* UNIT (add1 MAX-DISKS)))
(define RADIUS (/ UNIT 2))
```

```
(define ANIMATION-INTERVAL 1) ; ms
(define MOVE-DIST 2)
```

To enable scalability, we define the constant `UNIT` as a basic unit of measure (in pixels). By basing all the other measurements on this value, we can scale the entire interface by changing this one number. Constants `PEG-X` and `PEG-Y` are the location of the first peg. Constant `PEG-DIST` is the distance between pegs and `START-Y` is the *y* location of a disk in the bottommost position (position 1). Constant `RADIUS` is used to curve the ends of the disks and pegs. And constant `ANIMATION-INTERVAL` defines the time in milliseconds between animation updates, and `MOVE-DIST` defines how far a disk moves on each animation update. These last two parameters may need to be adjusted depending on the performance characteristics of the computer the code runs on.

Here are the main window elements. There should be no surprises here.

```
(define main-frame
  (new frame%
       [label "Tower of Hanoi"]
       [width (+ (* 7 UNIT) (* 3 PEG-DIST))]))

(define main-panel (new vertical-panel%
                        [parent main-frame]))

(define canvas
  (new canvas%
       [parent main-panel]
       [min-height (+ START-Y UNIT)]
       [paint-callback (λ (canvas dc) (update-canvas dc))]))

(define control-panel1
  (new horizontal-panel%
       [parent main-panel]))

(define control-panel2
  (new horizontal-panel%
       [parent main-panel]))
```

We'll examine `update-canvas` in greater detail later.

The following defines are for variables that will be updated during the progress of the animation.

```
(define num-disks 8)
(define delta-x 0)
(define delta-y 0)
(define target-x 0)
(define target-y 0)
(define source-peg 0)
(define dest-peg 0)
(define current-disk 0)
```

```
(define current-x 0)
(define current-y 0)
(define peg-disks (make-vector 3 0))
(define move-list '())
(define total-moves 0)
(define move-num 0)
(define in-motion #f)
(define mode 'stoppd)
```

The `peg-disks` variable is a three-element vector that's used to represent the three pegs. Each element of the vector will be populated with a list of numbers that represent the disks that reside on the respective pegs. Most of the other variables are fairly descriptive of their purpose, but we'll provide more detail as we progress through the section.

To give you an idea of what we're building, Figure 5-5 is a snapshot of the final application with a disk movement in progress.

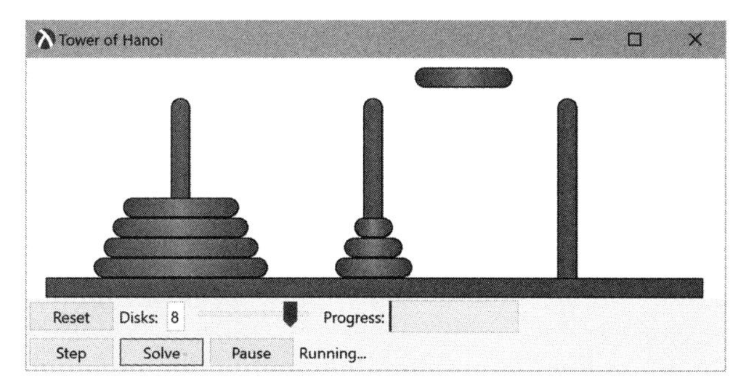

Figure 5-5: Tower of Hanoi GUI

In the next section, we'll describe the interface widgets that we'll use within the GUI.

Row 1 Widgets

The first row of controls (shown immediately below the canvas in Figure 5-5) is housed in the horizontal panel `control-panel1`. We'll describe each in the order they're added to the panel.

First, there's a basic button control that has a callback function that calls `reset` (provided the animation isn't already in progress), which restores everything back to the start state.

```
(define btn-reset
  (new button%
      [parent control-panel1]
      [label "Reset"]
      [callback (λ (button event)
                  (when (not in-motion) (reset)))]))
```

Next we have a text-field% widget. This is a basic text entry box. We use it to allow the user to specify the number of disks to use for the animation (via init-value, which is used to initialize a number of other controls as well).

```
(define text-disks
  (new text-field%
       [parent control-panel1]
       [label "Disks: "]
       [stretchable-width #f]
       [init-value "8"]))
```

Following the text box, we have a handy slider control. The slider provides an alternative method to select the number of disks. The callback function will update the text box containing the disk count based on the position of the slider. This is done by getting the value of the slider and sending it to the text box widget. The purpose of the remaining parameters should be fairly self-evident.

```
(define slider-disks
  (new slider%
       [parent control-panel1]
       [label #f]
       [stretchable-width #f]
       [min-width 100]
       [style (list' horizontal 'plain)]
       [min-value 1]
       [max-value MAX-DISKS]
       [init-value 8]
       [callback (λ (slider event)
                   (send text-disks
                         set-value
                         (number->string (send slider-disks get-value))))]))
```

The final element of the first row is a progress bar. Here we're using a Racket gauge% widget. To update this control, it just needs to be sent a number indicating the value to show (the number must be within the specified range).

```
(define gauge
  (new gauge%
       [parent control-panel1]
       [label "Progress: "]
       [stretchable-width #f]
       [min-width 100]
       (range 100)))
```

Row 2 Widgets

The second row of controls is housed in the horizontal panel control-panel2. First up is a button that allows the user to step through the solution of the puzzle one move at a time.

```
(new button%
     [parent control-panel2]
     [label "Step"]
     [callback (λ (button event)
                 (when (not in-motion)
                   (when (equal? move-list '()) (reset))
                   (set! in-motion #t)
                   (set! mode 'step)
                   (send msg set-label "Running...")
                   (init-next-step)
                   (send timer start ANIMATION-INTERVAL)))])
```

The state variable in-motion is used to flag whether an animation is currently in progress. If so, this variable is checked to ensure that the action isn't re-triggered before the step is complete. The move-list variable contains the list of moves provided by the solver, hanoi. If the list is empty, the solution has already been generated; in that case, the program is automatically reset. The variable mode can have one of three values:

'stopped. Waiting on user input.

'step. Performing a single step.

'solve. Solution animation in progress.

Then init-next-step sets up all the state variables to perform the next step of the solution.

Next we have the button that triggers the full-blown solution.

```
(new button%
     [parent control-panel2]
     [label "Solve"]
     [callback (λ (button event)
                 (when (not in-motion)
                   (let ([old num-disks]
                         [new (validate-disks)])
                 ❶ (when (or (equal? move-list '()) (not (= old new)))
                       (set! num-disks new)
                       (reset))
                     (set! in-motion #t)
                     (set! mode 'solve)
                     (send msg set-label "Running...")
                     (init-next-step)
                     (send timer start ANIMATION-INTERVAL))))])
```

The callback for this button is similar to the one for the step button, but this time we also check to see whether the user has changed the number of disks ❶ , in which case the program is reset to reflect the new number of disks before the animation is run. It also sets the mode variable to 'solve. (If you're bothered by the fact that (object:button . . .) prints out in the Dr-Racket window when this is executed, you can wrap this in a void form.)

The next button is quite simple: it sets mode to 'step. The mode is checked at the end of each step, so this will automatically cause the animation to stop. It can be resumed by pressing either the Solve or the Step button.

```
(new button%
     [parent control-panel2]
     [label "Pause"]
     [callback (λ (button event)
                  (set! mode 'step))])
```

The last control in the panel is a standard message% widget. It's used to display the current state of the program.

```
(define msg
  (new message%
       [parent control-panel2]
       [auto-resize #t]
       [label "Ready"]))
```

Getting in Position

Now let's look at a couple of functions that determine how to position the disks on the canvas. Each disk has a number from 1 to MAX-DISKS that determines the size of the disk. The disk-x function is passed the disk number and the number of the peg the disk resides on; it then returns the x-coordinate of the disk. The disk-y function is passed the disk position on a peg (see Figure 5-4) and returns the y-coordinate.

```
(define (disk-x n peg)
  (let ([w (* (add1 n) UNIT)])
    (- (+ PEG-X (* peg PEG-DIST)) (/ w 2) (/ UNIT -2))))

(define (disk-y pos)
  (- START-Y (* pos UNIT)))
```

A disk is actually just a rounded rectangle. We're going to add a little bling to our disks by including a gradient. This will make our disks look a little more disk-like. If you look closely at Figure 5-5, you'll notice that the disks are not a flat color, but give the appearance of being cylindrical. To this end we define a function make-gradient.

```
(define (make-gradient start stop c1 c2)
  (new linear-gradient%
       [x0 start] [y0 0]
       [x1 stop] [y1 0]
       [stops
        (list (list 0   (make-object color% c1))
              (list 0.5 (make-object color% c2))
              (list 1   (make-object color% c1)))]))
```

This function returns a Racket `linear-gradient%` object that can be applied to a brush. The `start` and `stop` parameters are the screen *x* locations where the gradient is to start and end. Variable `c1` is the color at the ends of the gradient, and `c2` is the color in the center of the gradient. Linear gradients work by defining a line segment whose end points are given by (x_0, y_0) and (x_1, y_1). The colors are applied perpendicularly along the line. That's why for our purposes we can just set the *y* values to 0. The colors are varied along the line by defining a list of *stop* positions. Each position defines the position along the line to apply a color. The position is a number between 0 and 1, where 0 is the color at the start of the line and 1 is the color at the end of the line; any number between 0 and 1 would designate a color at some point along the line.

To actually draw a disk, we use the following `draw-disk` function. We pass this function a drawing context, a disk number, a peg number, and a position on the peg.

```
(define (draw-disk dc n peg pos)
  (let* ([w (* (add1 n) UNIT)]
         [x (disk-x n peg)]
         [y (disk-y pos)])
    (send dc set-brush
          (new brush%
               [gradient (make-gradient x (+ x w) "Green" "GreenYellow")]))
    (send dc draw-rounded-rectangle x y w UNIT RADIUS)))
```

Inside we compute the width of the disk and the x-and y-coordinates. We then create a brush using our `make-gradient` function to create a linear gradient with which we draw a rounded rectangle to represent a disk.

The following functions draw the disks. The `draw-peg-disks` draws all the disks on a single peg. It's passed a drawing context, a peg number, and a list containing the disks that need to be drawn. The `draw-disks` function calls `draw-peg-disks` once for each peg.

```
(define (draw-peg-disks dc peg disks)
  (define (loop disks pos)
    (when (> pos 0)
      (let ([n (first disks)]
            [r (rest disks)])
        (draw-disk dc n peg pos)
```

```
      (loop r (sub1 pos)))))
  (loop disks (length disks)))

(define (draw-disks dc)
  (for ([peg (in-range 3)])
    (draw-peg-disks dc peg (vector-ref peg-disks peg))))
```

Here we have draw-base, which draws the base with the pegs (notice the pegs also use a linear gradient).

```
(define (draw-base dc)
  (for ([i (in-range 3)])
    (let ([x (+ PEG-X (* i PEG-DIST))])
      (send dc set-brush
            (new brush%
                 [gradient (make-gradient x (+ x UNIT) "Chocolate" "DarkOrange
    ")]))
      (send dc draw-rounded-rectangle x PEG-Y UNIT (+ UNIT (- START-Y PEG-Y))
    RADIUS)))
  (send dc set-brush (new brush% [color "Chocolate"]))
  (send dc draw-rectangle UNIT START-Y (+ (* 4 UNIT) (* 3 PEG-DIST)) UNIT))
```

The actual drawing process originates with update-canvas as given here. Aside from calling draw-base and draw-disks as described above, it checks to see if there's currently a disk in motion (determined by current-disk being greater than zero). If this is the case, it also renders the disk being animated.

```
(define (update-canvas dc)
  (draw-base dc)
  (draw-disks dc)
  (when (current-disk . > . 0)
    (let* ([w (* (add1 current-disk) UNIT)]
           [x current-x]
           [y current-y])
      (send dc set-brush
            (new brush%
                 [gradient (make-gradient x (+ x w) "Green" "GreenYellow")]))
      (send dc draw-rounded-rectangle x y w UNIT RADIUS))))
```

Controlling the Animation

We're now going to look at a couple of functions used to initialize the state variables at two key points in the process: at the start of the solution, and before each step. The first is called reset.

```
(define (reset)
❶ (set! num-disks (validate-disks))
  (set! delta-x 0)
  (set! delta-y 0)
```

```
    (set! current-disk 0)
  ❷ (set! move-list (hanoi num-disks 0 2))
    (set! total-moves (length move-list))
    (set! move-num 0)
  ❸ (vector-set! peg-disks 0 (range 1 (+ 1 num-disks)))
    (vector-set! peg-disks 1 '())
  ❹ (vector-set! peg-disks 2 '())
    (send canvas refresh-now))

❺ (define (init-next-step)
    (let ([move (first move-list)])
      (set! source-peg (first move))
      (set! dest-peg (second move))
      (set! delta-x 0)
      (set! delta-y (- MOVE-DIST))
      (set! target-y (/ UNIT 2))
      (set! move-list (rest move-list))
      (let* ([source-disks (vector-ref peg-disks source-peg)]
             [pos (length source-disks)])
        (set! current-disk (first source-disks))
        (set! current-x (disk-x current-disk source-peg))
        (set! current-y (disk-y pos))
        (vector-set! peg-disks source-peg
                     (rest source-disks)))))

❻ (define (validate-disks)
    (let* ([disks-str (send text-disks get-value)]
           [n (string->number disks-str)])
      (if (and (integer? n) (< 0 n (add1 MAX-DISKS)))
          (begin
            (send slider-disks set-value n)
            (send msg set-label "  Ready")
            n)
          (begin
            (send text-disks set-value (number->string num-disks))
            (send msg set-label "  Disks out of range.")
            num-disks))))
```

The code above begins by assigning the requested number of disks to the variable `num-disks` via a call to `validate-disks` ❶ (`validate-disks` ❻ ensures that the user has entered a proper number of disks; if not, it reverts back to the previous valid entry). Following this, a list of moves is generated based on the requested number of disks ❷. Next, we initialize the `peg-disks` vector with the disks on each peg ❸ ❹. The `init-next-step` function ❺ works by extracting the next move from the move list to determine the source and destination pegs, target locations, and parameters for the disk to be moved.

We finally arrive at the `move-disk` code responsible for the main animation update process. It processes a single move step in three phases: the target disk moving up and off the source peg, the disk moving left or right toward the destination peg, and the disk moving down to its final resting place. In the code below, update-progress is used to update the progress bar. It's called by move-disk at the end of each step ❺.

```
(define (update-progress)
  (send gauge set-value (inexact->exact (floor (* 100 (/ move-num total-moves)
    )))))

(define (move-disk)
  (cond [((abs delta-y) . > . 0)
         (begin
        ❶ (set! current-y (+ current-y delta-y))
           (when ((abs (- target-y current-y)) . < . MOVE-DIST)
             (set! current-y target-y)
             (if (delta-y . < . 0)
                 (begin ; was moving up
                   (set! target-x (disk-x current-disk dest-peg))
                   (set! delta-x (sgn (- dest-peg source-peg)))
                   (set! delta-y 0))
                 (begin ; was moving down
                   (set! move-num (add1 move-num))
                ❷ (vector-set! peg-disks dest-peg
                         (cons current-disk (vector-ref peg-disks dest-peg)))
                   (if (equal? mode 'step)
                       (begin
                         (send timer stop)
                         (set! current-disk 0)
                         (set! in-motion #f)
                         (set! mode 'stopped)
                         (send msg set-label "Ready")
                         (set! delta-y 0))
                       (if (> (length move-list) 0)
                           (init-next-step)
                           (begin
                             (send timer stop)
                             (send msg set-label "Done!")
                             (set mode 'stopped)
                             (set! in-motion #f)
                             (set! delta-y 0))))))))]

      ❸ [((abs delta-x) . > . 0)
         (begin
        ❹ (set! current-x (+ current-x delta-x))
           (when ((abs (- target-x current-x)) . < . MOVE-DIST)
             (set! current-x target-x)
             (set! target-y (* PEG-DIST (- dest-peg source-peg)))
             (set! delta-y MOVE-DIST)
```

```
          (let ([tdisks (length (vector-ref peg-disks dest-peg))])
            (set! target-y (disk-y (add1 tdisks))))
          (set! delta-x 0)))]

        [else (send timer stop)])

❺ (update-progress)
❻ (send canvas refresh-now))
```

Before move-disk is called, the target disk source and destination parameters would all have been established by a call to init-next-step.

The code for vertical movement starts ❶, and the current y-coordinate for the disk is updated. A check is then made to see whether the disk is at its target location. If the disk is at its target location and was moving up, a new x-coordinate target and travel delta for the disk are set. If the disk was moving down, the destination peg is updated by adding the animated disk to the list ❷. If the animation is in 'step mode, the animation timer is turned off and state variables are set to indicate the step is complete. Otherwise, the animation is in 'solve mode, so a check is made to see if any moves remain. If any moves remain, init-next-step is called; otherwise, the timer is stopped.

The check for horizontal movement is made at ❸. The x-coordinate of the disk is updated with a subsequent check to see whether the disk is at its target location ❹. If so, it sets up the state variables so that the next time move-disk is triggered, the disk begins to move down. At the end of every step, the progress meter and canvas are updated ❺ ❻.

Wrapping Things Up

The remaining items needed to get things going are given below. First, we define timer with a callback to move-disk. Next, reset is called to initialize all the state variables. And finally, we show the main window.

```
(define timer
    (new timer% [notify-callback move-disk]))

(reset)

(send main-frame show #t)
```

In addition to a nice little application to explore various facets of the Tower of Hanoi puzzle, in the process of building this application, we've utilized a significant number (but not all) of the widgets provided by the DrRacket environment.

Summary

In this chapter, we began with a couple of simple GUI applications and built up to a fairly full-featured application where we were exposed to a number of the elements needed to build a robust application. Later in the book, we'll see how to bundle this functionality into a stand-alone application that can be run independent of the DrRacket environment. But, next up, we'll be exploring how to access and analyze data in its various incarnations.

6

DATA

This chapter is all about data: reading it, writing it, visualizing it, and analyzing it. We'll begin with a discussion of input and output using Racket ports (an essential tool we'll use throughout this chapter).

I/O, I/O, It's Off to Work We Go

Data is transferred in Racket via ports: data flows into an input port and out of output ports. The general process of transferring data to or from an external source (such a text file or database server) is referred to as *I/O*. Let's look at some examples of ports.

File I/O Ports

Ports can be used for writing and reading data to a file, as the following dialog illustrates.

```
> ; Output some stuff
(define out-port (open-output-file "data/SomeStuff.txt"))
(display "some stuff" out-port)
(close-output-port out-port)
```

```
> ; Read it back in
(define in-port (open-input-file "data/SomeStuff.txt"))
(read-line in-port)
(close-input-port in-port)
"some stuff"
```

If we attempt to open a port to a file that already exists, we get an error.

```
> (define more-out (open-output-file "data/SomeStuff.txt"))
open-output-file: file exists
  path: ...\data\SomeStuff.txt
```

The default mode of operation for `open-output-file` is to create a new file. Since we can't create a new file twice, we need to declare how we intend to handle an existing file. This is managed via the `#:exists` keyword: we can append to the existing file by specifying `'append` as its value, blow away the file and create a new file with `'replace`, or keep the file but delete the contents with `'truncate` (the default for value for `#:exists` is `'error`, which means to generate an error if the file exists). We exercise a few of these options below.

```
> (define out-port (open-output-file "data/SomeStuff.txt" #:exists 'append))
  (display "some more stuff\n" out-port)
  (close-output-port out-port)

> (define in-port (open-input-file "data/SomeStuff.txt"))
> (read-line in-port)
"some stuff"

> (read-line in-port)
"some more stuff"
> (close-input-port in-port)

> (define out-port (open-output-file "data/SomeStuff.txt" #:exists 'truncate))
  (display "some new stuff\n" out-port)
  (close-output-port out-port)

> (define in-port (open-input-file "data/SomeStuff.txt"))
> (read-line in-port)
"some new stuff"
> (read-line in-port)
#<eof>
> (close-input-port in-port)
```

Once the end of the file is reached, `read-line` returns an end of file object; this prints as `#<eof>` but is defined as `eof` in Racket. This value can be tested for with the `eof-object?` predicate.

```
> eof
#<eof>

> (eof-object? eof)
#t
```

Every time a port is opened, you must remember to close it when the data transfer is complete. The closing can be performed automatically by using either `call-with-output-file` or `call-with-input-file` (depending on which way the data is flowing). These procedures work by supplying them with a function that does the actual data transfer. Here are some examples using this approach.

```
> (call-with-output-file "data/SomeData.txt"
    #:exists 'truncate
    (λ (out-port)
      (display "Data line1\n" out-port)
      (display "Data line2\n" out-port)
      (display "Data line3\n" out-port)
      (display "Data line4\n" out-port)))

> (call-with-input-file "data/SomeData.txt"
    (λ (in-port)
      (let loop()
        (let ([data (read-line in-port)])
          (unless (eof-object? data)
            (displayln data)
            (loop))))))
Data line1
Data line2
Data line3
Data line4
```

We'll explore file ports in greater detail a bit later in the chapter.

String Ports

Ports can be opened against strings. This can be handy when trying to build a string where different components of the string will be appended at various times within a procedure. We'll put string ports to good use in Chapter 10 when we build string representations of algebraic expressions. Some simple examples are provided here:

```
> (define str-port (open-output-string))
> (display "Hello "   str-port)
  (display "there, " str-port)
  (display "amigo!"   str-port)
```

```
> (get-output-string str-port)
"Hello there, amigo!"
```

Unlike file ports, string ports don't have to be explicitly closed.

Computer-to-Computer Ports

Ports can be set up to allow two computers to communicate with each other. This type of communication uses the TCP/IP protocol. To establish the connections, the first computer (called the *server*) establishes itself as a listener with the `tcp-listen` command. This command accepts an unsigned integer as a port number (this is a TCP/IP port number). Note that a Racket port is a different entity from the TCP port specified by the port number. The server then calls `tcp-accept`, which returns two values—an input port and an output port—to permit two-way communication between the computers. The following session illustrates setting up a server and the server waiting for a query from a client computer.

Computer 1 – The Server

```
> (define comp1 (tcp-listen 999))
  (define-values (comp1-in comp1-out) (tcp-accept comp1))
  (read-line comp1-in)
  (displayln "Got it, bro!\n" comp1-out)
  (close-input-port comp1-in)
  (close-output-port comp1-out)
 "Hello there!"
```

The string `"Hello there!"` is sent from the client. This is the result of executing the line `(read-line comp1-in)`, after which the server responds with `"Got it, bro!"`.

The client establishes its communication link by a using `tcp-connect`. The `tcp-connect` command takes the server computer name and the port number established by the server as arguments. It then initiates the dialog by sending `(displayln "Hello there!\n" comp2-out)` over the output port and waiting for a response from the server with `(read-line comp2-in)`.

Computer 2 – The Client

```
> (define-values (comp2-in comp2-out) (tcp-connect "Comp1Name" 999))
  (displayln "Hello there!\n" comp2-out)
  (flush-output comp2-out)
  (read-line comp2-in)
  (close-input-port comp2-in)
  (close-output-port comp2-out)
"Got it, bro!"
```

This is just a simple example; there are of course a number of nuances to setting up successful communication channels between computers. Consult the Racket Documentation for additional details.

Introduction to Security

Now that we've explored ports, let's see how we can use them to enhance our security. No, not computer security. In this section, we'll take a look at a type of security that virtually everyone is interested in: money. In particular we'll be exploring securities such as stocks and bonds. The specific data we're going to look at is price over time. There are a number of ways to look at prices: stock prices of individual corporations, average prices over an industry or group of institutions, or prices of index funds.

Table 6-1 lists the various entities we'll be investigating. In the remainder of the chapter, we'll refer to these entities as *assets*. The Symbol column indicates the stock market symbol that's used to look up price information. Definitions for entries in the Type column are as follows:

Corp An individual corporation. Most of these should be reasonably familiar.

Index A market index—just an indicator and not something you actually invest in. The Dow Jones Industrial Average is the weighted average consisting of 30 large publicly traded institutions. The S&P 500 (Standard and Poor's 500) is similar but consists of 500 large publicly traded institutions.

Index Fund An index fund, unlike an index, *can* be invested in. An index fund is typically composed of some mix of stocks or bonds. The Vanguard Total Bond Market fund is composed of a mix of long-term and short-term and both corporate and government bonds. The idea of index funds is that investing in a range of institutions minimized risk since the poor performance of one investment will be mitigated when averaged in with the remaining investments.

Table 6-1: A Selection of Securities

Name	Symbol	Type
Amazon	AMZN	Corp
Apple	AAPL	Corp
Bank of America	BAC	Corp
Dow Jones	^DJI	Index
ExxonMobil	XOM	Corp
Ford	F	Corp
Microsoft	MSFT	Corp
S&P 500	^GSPC	Index
Vanguard Total Bond Market	VBMFX	Index Fund

The data we're going to examine was downloaded from the Yahoo! Finance website. This data is in the form of *comma-separated value (CSV)* files. What this means is that each value in the file is separated by a comma, and each record occupies a single line. Here's an example from a file containing the Dow Jones Industrial Average for the first few days of the year 2007 (we've lopped off a few decimal places from the prices to keep the listing from extending off the page).

```
Date,Open,High,Low,Close,Adj Close,Volume
2007-01-03,12459.540,12580.349,12404.820,12474.519,12474.519,327200000
2007-01-04,12473.160,12510.410,12403.860,12480.690,12480.690,259060000
2007-01-05,12480.049,12480.129,12365.410,12398.009,12398.009,235220000
```

The first row gives a short description of each data value in the succeeding rows. Table 6-2 gives a more detailed description.

Table 6-2: CSV File Format

Value	Description
Date	Date of transactions (year-month-day)
Open	Price when market opened
High	Highest traded price for the day
Low	Lowest traded price for the day
Close	Price when the market closed
Adj Close	Adjusted closing price
Volume	Number of trades for the day

The adjusted close reflects any adjustments based on dividends or stock splits (the term *split* means if you owned a single stock share selling for $100, you'd own two shares at $50 after the split).

Getting Data into Racket

The first order of business is to get the CSV data into a form that will be useful within Racket. We'll use a `hist` structure to contain the data for a single record from the CSV file. This struct has the following form.

```
(struct hist (date open high low close adj-close vol)
  #:transparent)
```

The field names should need no explanation.

The following function will take a file port and return a `hist` structure populated with the data values from the current import record (the next unread line in the file) with each price entry converted to a numeric value.

```
(define (import-record port)
  (let ([rec (read-line port)])
    (if (eof-object? rec)
        eof
      ❶ (match (string-split rec ",")
        ❷ [(list date open high low close adj-close vol)
         ❸ (hist date
                 (string->number open)
                 (string->number high)
```

```
                (string->number low)
                (string->number close)
                (string->number adj-close)
                (string->number vol))]
      ❹ [_ (error "Failed to load record.")]))))
```

Here we've taken the opportunity to exercise another of Racket's hidden treasures, pattern matching. *Pattern matching* uses the match form ❶ included in the *racket/match* library (not to worry; this library is automatically included with the *racket* library).

A match expression looks a bit like a cond expression, but instead of having to use a complex Boolean expression, we simply provide the data structure we want to match against. It's possible to use a number of different structures as patterns to match against, including literal values, but we'll simply use a list for this exercise ❷. The split rec values are bound to the identifiers date, open, and so on. If a match is found, then a hist structure is returned ❸. A single underscore (_) serves as a wildcard that matches anything ❹ (for example if the number of values in the split list did not match the number of items in the binding list, an error exception would be raised). With this in place, the following code reads in a few values (in the following code segments, the reader should substitute their own data path where paths starting with StockHistory/ are used):

```
> (define in-port
    (open-input-file "StockHistory/Daily/XOM.csv"))
> (import-record in-port)
(hist "Date" #f #f #f #f #f #f)

> (import-record in-port)
(hist "1980-01-02" 3.445313 3.453125 3.351563 3.367188 0.692578 6622400)

> (import-record in-port)
(hist "1980-01-03" 3.320313 3.320313 3.25 3.28125 0.674902 7222400)

> (close-input-port in-port)
```

The first hist structure produced contains false values because the header strings in the first line of the CSV file couldn't be converted to numbers by import-record.

For display purposes, we often want to uniformly format the history records to display each value as a string right-aligned and with a certain precision. The following function performs this service:

```
(define (format-rec rec width prec)
  (match rec
    [(hist date open high low close adj-close vol)
     (hist date
           (~r open #:min-width width  #:precision (list '= prec))
           (~r high #:min-width width  #:precision (list '= prec))
```

```
        (~r low #:min-width width  #:precision (list '= prec))
        (~r close #:min-width width  #:precision (list '= prec))
        (~r adj-close #:min-width width  #:precision (list '= prec))
        (~r vol #:min-width 9 ))]))
```

The width parameter specifies the overall width of each value, and the prec parameter specifies the precision. Applying this function to the first few rows of the ExxonMobil data results in the following:

```
> (define in-port
    (open-input-file "StockHistory/Daily/XOM.csv"))
> (import-record in-port)
(hist "Date" #f #f #f #f #f #f)

> (format-rec (import-record in-port) 6 2)
(hist "1980-01-02" "  3.45" "  3.45" "  3.35" "  3.37" "  0.69" "  6622400")

> (format-rec (import-record in-port) 6 2)
(hist "1980-01-03" "  3.32" "  3.32" "  3.25" "  3.28" "  0.67" "  7222400")

> (close-input-port in-port)
```

Notice that some of the output values are padded with spaces because the numbers are less than six characters wide when formatted with two decimal places.

Since we'll sometimes want to display the data in tabular form, we'll take advantage of the text-table package. Unfortunately, this package isn't included with the default Racket install, so it'll need to be installed via either Racket's package manager or the raco command line tool (see Chapter 2 on infix for an example of how to install packages). Once the text-table package is installed, the following command must be included in your definition file or executed in the interactive window:

```
(require text-table)
```

The text-table package defines the table->string function, which takes a list of lists, where each sublist represents a row in the table. Here's a simple example of how it can be used.

```
> (define data '((a b c) (1 2 3) (4 5 6)))
> (display (table->string data))
+-+-+-+
|a|b|c|
+-+-+-+
|1|2|3|
+-+-+-+
|4|5|6|
+-+-+-+
```

The data that we're going to be querying against has a file structure similar to the one shown in Figure 6-1. The files have records (one record for each day, month, or week) for years 1980 through the end of 2016 (where available).

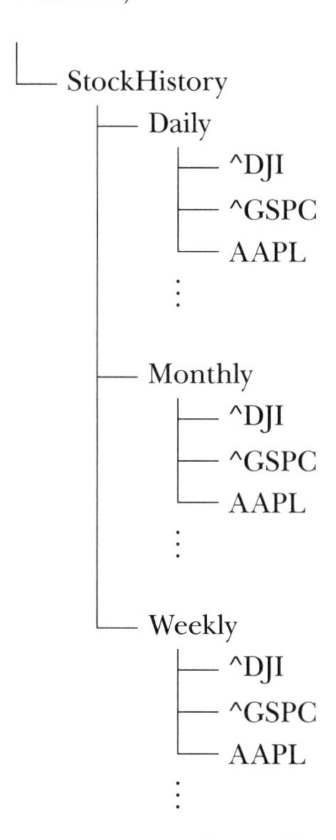

Figure 6-1: Stock history file structure

The following code will display rows of stock history when given the stock symbol and time period (either `"Daily"`, `"Monthly"`, or `"Weekly"`—files for each period are stored in a corresponding folder). In addition, a filter must be specified. The filter is a function that accepts a `hist` structure and returns `#t` or `#f` depending on what's being searched for.

```
(define (show sym period filter)
  (let ([in-port (open-input-file
      (build-path "StockHistory" period (string-append sym ".csv")))])
    (read-line in-port) ; skip past header row
 ➊ (let* ([recs
          (reverse
          ➋ (let loop([rec-list '()])
            ➌ (let ([rec (import-record in-port)])
                (if (eof-object? rec)
                    rec-list
                  ➍ (if (filter rec)
                      ➎ (let* ([rec (format-rec rec 8 2)]
                                [rec (list (hist-date rec)
```

```
                                          (hist-high rec)
                                          (hist-low rec)
                                          (hist-close rec))])
                           ❻ (loop (cons rec rec-list)))
                         ❼ (loop rec-list))))))]
        ❽ [tbl-list (cons (list "   Date" "   High" "   Low" "   Close") recs)])
      (close-input-port in-port)
    ❾ (display (table->string tbl-list)))))
```

Once the input file is opened and we've skipped past the header line, the code line (let* ([recs ❶ binds two variables: recs and tbl-list ❽. Note that recs is used to initialize tbl-list ❽, which simply adds a header for the data contained in recs. Then the final output is generated ❾.

In the main body of code, we set up a function called loop ❷ to recursively extract the data from the input file (the reverse just prior to this is needed since the recursive call builds up the data in the reverse order). Note that rec-list is initialized with an empty list. The identifier rec ❸ is populated with a single row from the input file. Once the end of the file is reached, we output the compiled rec-list, but until then we use the filter to search for records that meet our criteria ❹. When such a record is found, we bind a local version of rec using our previously defined format-rec code ❺. Since we're in a let* form, we bind a new local version of rec on the next line with data extracted from the imported record. Having done this, we add this to the previously imported data ❻ and recursively call loop. If the filter's criteria are not met, we simply trigger the reading of the next row of data from the import file with the existing data ❼.

Let's take a look at the year 2008 when things started to go horribly wrong in the financial sector. (The substring function is used to pull the year out of the date field and display only records where the year is equal to 2008.)

```
> (show "^DJI" "Monthly"
        (λ (rec) (equal? (substring (hist-date rec) 0 4) "2008")))

+----------+--------+--------+--------+
|   Date   |  High  |  Low   |  Close |
+----------+--------+--------+--------+
|2008-01-01|13279.54|11634.82|12650.36|
+----------+--------+--------+--------+
|2008-02-01|12767.74|12069.47|12266.39|
+----------+--------+--------+--------+
|2008-03-01|12622.07|11731.60|12262.89|
+----------+--------+--------+--------+
|2008-04-01|13010.00|12266.47|12820.13|
+----------+--------+--------+--------+
|2008-05-01|13136.69|12442.59|12638.32|
+----------+--------+--------+--------+
|2008-06-01|12638.08|11287.56|11350.01|
```

```
+----------+--------+--------+--------+
|2008-07-01|11698.17|10827.71|11378.02|
+----------+--------+--------+--------+
|2008-08-01|11867.11|11221.53|11543.96|
+----------+--------+--------+--------+
|2008-09-01|11790.17|10365.45|10850.66|
+----------+--------+--------+--------+
|2008-10-01|10882.52| 7882.51| 9325.01|
+----------+--------+--------+--------+
|2008-11-01| 9653.95| 7449.38| 8829.04|
+----------+--------+--------+--------+
|2008-12-01| 9026.41| 8118.50| 8776.39|
+----------+--------+--------+--------+
```

During that year, the Dow swung from a high of 13,279.54 in January to a low of 7,449.38 in November. A swing of 5,380 points, or a 44 percent drop!

A year earlier, Microsoft had been averaging around $30 a share, but it dropped below $19 a few times in 2008. Let's see when that happened.

```
> (show "MSFT" "Daily"
    (λ (rec) (and
                  (< (hist-close rec) 19)
                  (equal? (substring (hist-date rec) 0 4) "2008"))))

+----------+--------+--------+--------+
|   Date   |  High  |  Low   | Close  |
+----------+--------+--------+--------+
|2008-11-19|   19.95|   18.25|   18.29|
+----------+--------+--------+--------+
|2008-11-20|   18.84|   17.50|   17.53|
+----------+--------+--------+--------+
|2008-12-01|   19.95|   18.60|   18.61|
+----------+--------+--------+--------+
|2008-12-29|   19.21|   18.64|   18.96|
+----------+--------+--------+--------+
```

A Database Detour

This section is entirely optional since the rest of the chapter isn't dependent on the ideas presented here. But we do want to present some information on random file access that could be useful in various scenarios. *Random file access* is a key component for efficient data queries. In particular, random file access is a critical component in any database. A database is mainly just a predefined collection of tables, with each table potentially containing multiple records. So far, to locate a particular record in a table, we've scanned a file, record by record, until the desired entry was located. On average half the records have to be checked, and in the worst case, all the records are checked. This is clearly not very efficient.

Normally, we're looking for a particular date, and the fact that the records are in date order suggests that a binary search might be applicable (if you don't know what a binary search is, we'll delve into the details shortly); but there is a problem. In CSV files, the records are packed together with a variable number of characters in each record, so there is no way to accurately get positioned on a particular record without reading through the file starting from the beginning. The key is to allocate slots for each record where each slot is a fixed size, but large enough to hold the largest record. To facilitate this, we'll define the following function:

```
(define (file-info sym period)
  (let ([port (open-input-file
          (build-path "StockHistory" period (string-append sym ".csv")))]
        [recs 0]
        [max-size 0])
    (let loop ()
      (let ([rec (read-line port)])
        (unless (eof-object? rec)
          (let ([len (string-length rec)])
            (set! recs (add1 recs))
            (unless (<= len max-size)
              (set! max-size len)))
          (loop))))
    (close-input-port port)
    (values recs max-size)))
```

This function scans the input file to determine the number of records in the file and the maximum record size.

With the information returned from `file-info`, we can construct an appropriately formatted data file. This file reserves the first slot to hold the information returned from `file-info` since this will be useful when we get around to actually searching the file. The remaining slots are populated with values from the source file by using the `file-position` function to set the location the values are written to (by calling `display`). Unused space at the end of a record is filled with zero (0) bytes. To actually create the file, we define a function called `csv->db`:

```
(define (csv->db sym period)
  (let*-values ([(recs max-size) (file-info sym period)]
      [(in-port) (open-input-file
        (build-path "StockHistory" period (string-append sym ".csv")))]
   ➊ [(out-port) (open-output-file
        (build-path "StockHistory" period (string-append sym ".db"))
          #:exists 'truncate)]
      [(slot-size) (+ 10 max-size)])
 ➋ (file-position out-port (* recs slot-size))
 ➌ (display (make-string slot-size #\space) out-port)
```

```
    (file-position out-port 0)
❹  (display recs out-port)
    (display "," out-port)
❺  (displayln slot-size out-port)
    (read-line in-port) ; read past header
    (for ([ i (in-range 1 recs)])
      (let ([rec (read-line in-port)]
            [pos (* i slot-size)])
     ❻ (file-position out-port pos)
     ❼ (displayln rec out-port)))
    (close-input-port in-port)
    (close-output-port out-port)
    ))
```

This function creates a data file in the same folder as the source file, with the same name, except that the output file ❶ has *.db* as a file extension instead of *.csv*. Next, we set the file position to the end of the file ❷ and write a dummy record ❸. This is an efficiency step to keep the operating system from constantly enlarging the file every time a record is written. We write the number of records and slot size to the first record in the file ❹ ❺. For each record in the input file, we then position the output file pointer ❻ and write out the record ❼.

Having thus created a searchable data file, we can now create a routine to perform a binary search to efficiently find a record for a particular date. (For more detailed information on binary searches in general, see the Wikipedia article: *http://en.wikipedia.org/wiki/Binary_search_algorithm*.)

```
(define (bin-search sym period date)
❶ (let* ([port (open-input-file
              (build-path "StockHistory" period (string-append sym ".db")))]
         [info (string-split (read-line port) ",")]
         [recs (string->number (first info))]
         [slot-size (string->number (second info))]
         [min 1]
      ❷ [max recs])

  ❸ (define (get-date rec)
       (substring rec 0 10))

  ❹ (define (get-rec i)
       (file-position port (* slot-size i))
       (read-line port))

    (let loop ()
   ❺ (if (> min max)
          (begin
            (close-input-port port)
            #f)
```

```
❺ (let* ([i (floor (/ (+ min max) 2))]
         [rec (get-rec i)]
         [d (get-date rec)])
    (cond [(string<? d date)
        ❻ (set! min (add1 i))
           (loop)]
          [(string>? d date)
        ❼ (set! max (sub1 i))
           (loop)]
          [else
           (close-input-port port)
        ❽ rec]))))))
```

First, we perform some basic initialization ❶ ❷. After opening the input file, we read the first record (which contains the number of records and slot size) and bind `recs` and `slot-size` to the appropriate values. Next, we define a couple of helper functions to simplify retrieving data for the current record ❸ ❹. The remainder of the code is a straightforward implementation of a binary search routine, the main body consisting of a loop (via recursive function `loop`). The search begins by testing whether any records remain to be checked ❺. If `(> min max)` is true, all records have been checked and no match has been found, so the function returns `#f`. Next, we compare the date of the middle element of the file with the target date ❻. If the target date matches the middle element, the current record is returned ❽. If the target date is less than or greater than the date of the current element, the search continues by narrowing down the range of the file by resetting `min` ❻ and `max` ❼ as needed.

We test our creation by first forming a searchable data file for Microsoft stock prices with the following:

```
> (csv->db "MSFT" "Daily")
```

If we're then interested in retrieving the stock price record for March 13, 1992, we do it like this:

```
> (bin-search "MSFT" "Daily" "1992-03-13")
"1992-03-13,2.552083,2.562500,2.510417,2.520833,1.674268,36761600"
```

The Microsoft file has 7,768 records. On average, a linear search would have to examine 3,884 records. The worst-case performance of a binary search is given by the following expression, where t is the number of checks that need to be performed and n is the number of records:

$$t = \lfloor \log_2(n + 1) \rfloor$$

This means that searching the Microsoft data would only require the following checks. Examining 12 records is a lot better than examining 3,884.

$$\begin{aligned} t &= \lfloor \log_2(7768 + 1) \rfloor \\ &= 12 \end{aligned}$$

Data Visualization

A picture is worth a thousand dollars, er . . . a thousand words. So far we've been looking at data. Now we want to look at information. The difference is this: data is just a raw assemblage of numbers, dates, and strings; information says how these things are related. Just perusing a list of numbers doesn't provide much insight into their meaning, but often a visual representation invokes an epiphany. In light of this, we turn to the topic of data visualization. In this section we'll look at financial data in two different ways: value over time and frequency analysis via histograms.

To ensure that we have everything we need, we start from scratch with the following definitions.

```
#lang racket
(require plot)
(require text-table)

(define (data-path symbol period)
  (build-path "StockHistory" period (string-append symbol ".csv")))

(struct hist (date open high low close adj-close vol)
  #:transparent)

(define symbols '("^DJI" "^GSPC" "AAPL" "AMZN"
                        "BAC" "F" "MSFT" "VBMFX" "XOM"))

(define symbol-color
  (make-hash
   (list
    (cons "^DJI" "black")
    (cons "^GSPC" "gray")
    (cons "AAPL" "black")
    (cons "AMZN" "gray")
    (cons "BAC" "purple")
    (cons "F" "orange")
    (cons "MSFT" "blue")
    (cons "VBMFX" "black")
    (cons "XOM" "gray")
    )))

(define symbol-style
  (make-hash
   (list
    (cons "^DJI" 'solid)
    (cons "^GSPC" 'solid)
    (cons "AAPL" 'dot)
    (cons "AMZN" 'dot)
    (cons "BAC" 'dot-dash)
```

```
    (cons "F" 'solid)
    (cons "MSFT" 'long-dash)
    (cons "VBMFX" 'short-dash)
    (cons "XOM" 'short-dash)
    )))
```

Most of this should be self-explanatory. As a reminder, the `text-table` package isn't part of the default Racket setup (see "Getting Data into Racket" on page 150 for more info). We use `data-path` to keep from hardcoding the file paths in the body of various functions. To differentiate multiple assets on plots, they're assigned unique colors in `symbol-color` and line styles in `symbol-style`.

We'll again make use of `import-record`, which is reproduced here.

```
(define (import-record port)
  (let ([rec (read-line port)])
    (if (eof-object? rec)
        eof
        (match (string-split rec ",")
          [(list date open high low close adj-close vol)
           (hist date
                 (string->number open)
                 (string->number high)
                 (string->number low)
                 (string->number close)
                 (string->number adj-close)
                 (string->number vol))]
          [_ (error "Failed to load record.")]))))
```

While this function extracts everything in the record, for the remainder of the chapter we'll primarily be interested in the date and closing price.

Since we'll be plotting time-varying values, we need to convert the date string from each record into a numeric value. We can do this with the following vector and function:

```
(define month-days
  #(0 0 31 59.25 90.25 120.25 151.25 181.25
      212.25 243.25 273.25 304.25 334.25))

(define (date->number d)
  (match (string-split d "-")
    [(list year month day)
     (let ([year (string->number year)]
           [month (string->number month)]
           [day (string->number day)])
       (exact->inexact (+ year
                          (/ (vector-ref month-days month) 365.25)
                          (/ (sub1 day) 365.25))))]))
```

The month-days vector provides the number of elapsed days in the year for the first day of any given month (the month number is the index; for example, February is represented by index 2). For example the entry with index 2 is 31, indicating that on February 1, 31 days have elapsed. Months after February have an extra quarter-day to account for leap years. The date conversion happens in date->number where the year forms the integer portion of the date, and the month and day provide the fractional portion. The fact that February can have 28 or 29 days is handled by an approximation that should be sufficient for our purpose.

Plotting for Success

As you saw in the Chapter 4, Racket plots a sequence of line segments by using a lines form that takes a list of vectors as its argument. Each vector specifies the *x* (date) and *y* (close price) coordinates of a segment end point. To construct this list, we use the get-coords function:

```
(define (get-coords symbol period filter normalize)
  (let ([in-port (open-input-file (data-path symbol period))]
        [start-price #f])
    (read-line in-port)
    (let* ([recs
      (reverse
        (let loop([rec-list '()])
          (let ([rec (import-record in-port)])
            (if (eof-object? rec)
               rec-list
               (if (filter rec)
                 (let ([date-val (date->number (hist-date rec))]
                     [close (hist-close rec)])
                   (unless start-price (set! start-price close))
                   (let ([val
                 ➊ (if normalize (/ close start-price) close)])
                     (loop
                       (cons (vector date-val val) rec-list))))
                 (loop rec-list))))))])
      (close-input-port in-port)
      recs)))
```

As with other examples, this function takes the stock symbol, time period type, and a filter function as arguments. Since the assets we're examining can have widely differing values, to enable showing them on the same plot we provide the ability to normalize the values with an extra parameter. What this means is that instead of plotting the actual value, we plot the ratio of the first value in the time period to the actual value ➊. This way all the assets start out with a value of 1, but we still see how the relative values vary over time. A few examples will make this clear shortly.

The routine that does the plotting is quite simple.

```
(define (plot-symbols symbols period filter
                      [normalize #f]
                      [anchor 'top-left]
                      [y-min #f]
                      [y-max #f])
  (let* ([plot-data
          (for/list ([symbol symbols])
            (let ([color (hash-ref symbol-color symbol)]
                  [style (hash-ref symbol-style symbol)])
              (lines (get-coords symbol period filter normalize)
                     #:label symbol
                     #:width 1.5
                     #:color color
                     #:style style)))]
         [ymin (if (and normalize (not y-min)) 0.0 y-min)]
         [ymax (if (and normalize (not y-max)) 2.0 y-max)])
    (parameterize
        ([plot-width 400]
         [plot-height 250]
         [plot-x-label "Year"]
         [plot-y-label #f]
         [plot-legend-anchor anchor])
      (plot plot-data
            #:y-min ymin
            #:y-max ymax))))
```

This time we provide it with a list of stock symbols (to allow multiple assets to be plotted at the same time), a time period type, and a filter function as arguments. We also have the option to specify whether or not the plot normalizes the data with the optional `normalize` parameter, which defaults to `#f`. Since plotted values can occur on just about any portion of the plot, we allow the user to specify where the legend is positioned with the optional `anchor` parameter. In addition we allow the user to override the default range of *y* values.

Since we're primarily going to be plotting data for some range of years or dates, we'll define a couple of function factories that create query functions by specifying the range of dates we're interested in.

```
(define (year-range y1 y2)
  (λ (rec)
    (string<=? y1 (substring (hist-date rec) 0 4)
               y2)))

(define (date-range d1 d2)
  (λ (rec)
    (string<=? d1 (substring (hist-date rec) 0 10)
               d2)))
```

Having established the prerequisites, we're ready to generate some plots. Let's begin by plotting Dow Jones data for 2007 and 2008 (see Figure 6-2).

```
> (plot-symbols '("^DJI") "Daily"
                (year-range "2007" "2008")
                #f 'bottom-left)
```

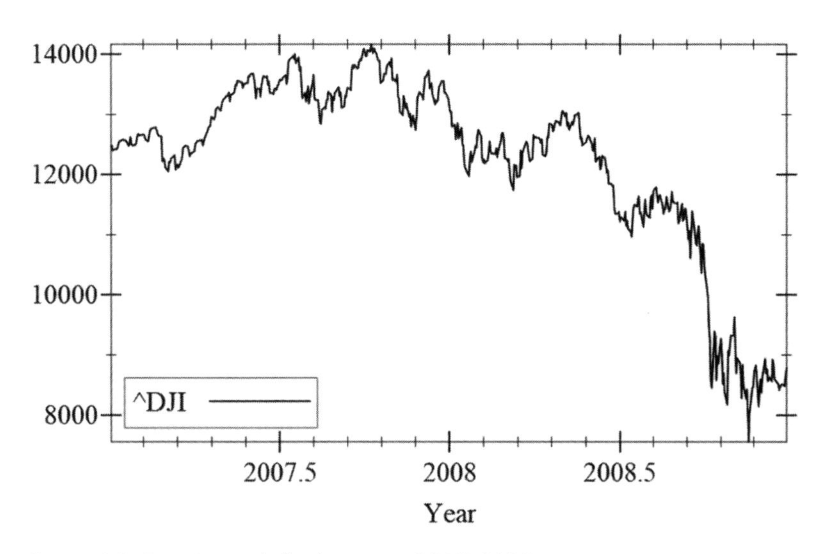

Figure 6-2: Dow Jones daily close price 2007–2008

Now we can actually *see* the precipitous drop that occurred in October of 2008. Next, let's look at a few other institutions to see how they fared (see Figure 6-3).

```
> (plot-symbols '("^DJI" "^GSPC" "AAPL" "VBMFX") "Daily"
                (year-range "2007" "2008")
                #f 'bottom-left)
```

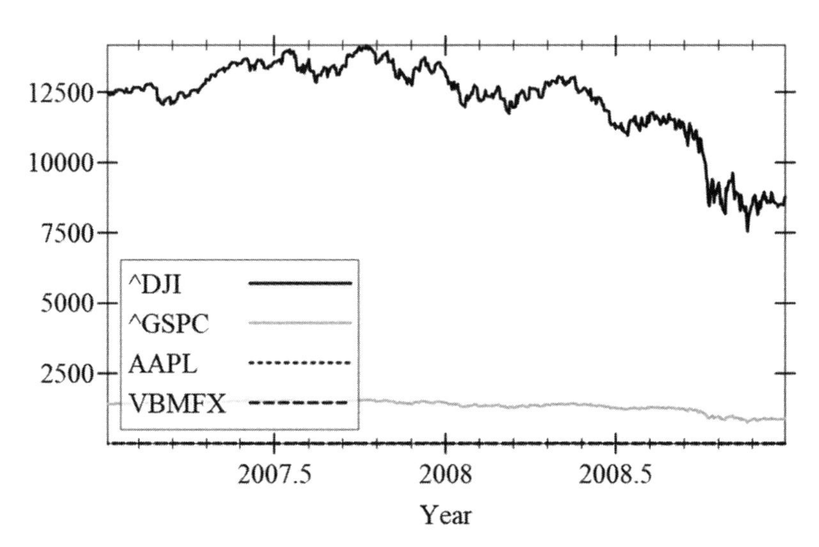

Figure 6-3: Multiple daily close prices

Unfortunately, the Dow numbers are so large, they've swamped everybody else. Now let's see what happens when we normalize the numbers (see Figure 6-4).

```
> (plot-symbols '("^DJI" "^GSPC" "AAPL" "VBMFX") "Daily"
                (year-range "2007" "2008")
                #t 'bottom-left)
```

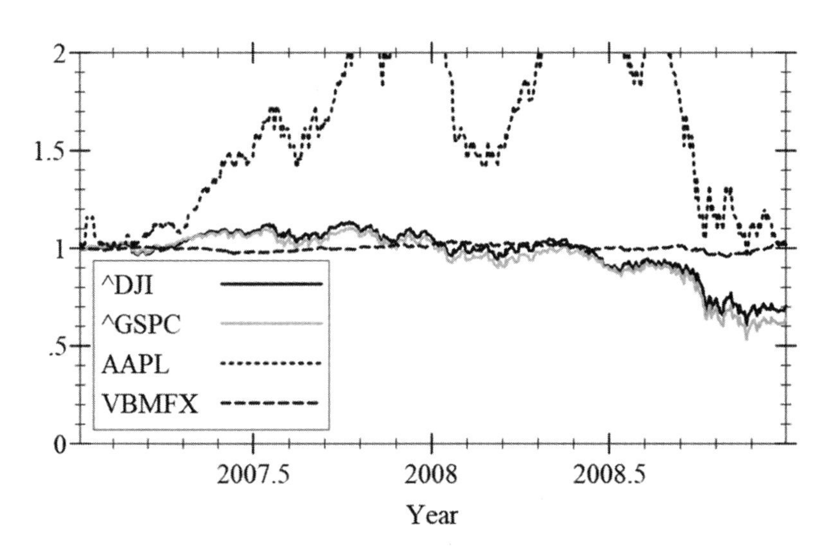

Figure 6-4: Normalized daily close prices

It's clear that the Dow and S&P 500 general market indices track fairly close to each other. Apple is all over the map. The bond fund stayed steady throughout the mayhem.

What have we learned? Well, it would be foolish to make too many assumptions based on a few plots. It's fairly clear, though, that at least for this time period, bonds barely wiggled, the market as a whole (represented by the index indicators S&P 500 and the Dow) had some turbulence, but not as much as Apple.

Let's take a longer view and see what comes up in Figure 6-5.

```
> (plot-symbols '("^GSPC" "AAPL" "XOM") "Monthly"
                (year-range "1981" "2016")
                #t 'top-left 0 20)
```

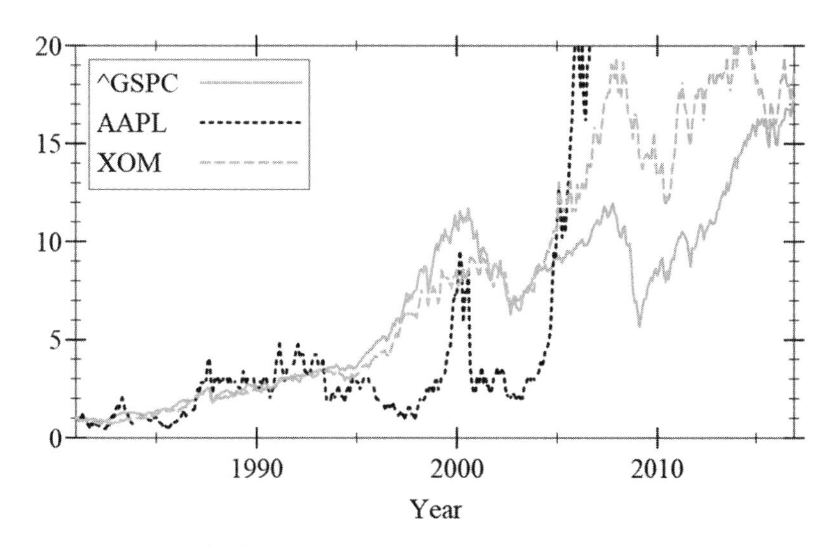

Figure 6-5: Monthly close prices 1981–2016

So around 2005-ish, Apple went off the chart (more about that a bit latter). Let's examine ExxonMobil in a bit more detail. Since this is a normalized chart, we're seeing the relative difference in the prices; this means that at the end of 2016, ExxonMobil stock was worth roughly 18 times what it was worth at the beginning of 1981. Sounds like a lot, but is it really? We can get some idea of the average annual return this represents by using the compound interest formula given by the following:

$$V = P\left(1 + \frac{i}{n}\right)^{nt}$$

In this formula, V is the current value, P is the initial principal, i is the annual interest rate, n is the number of compounding periods per year (we're going to assume monthly compounding, so this will be 12), and t is the number of periods (so nt is 12 times the number of years). We want to know the annual interest i, so after a bit of algebra (we won't bore you with the details), we get this:

$$i = n\left(\left(\frac{V}{P}\right)^{\frac{1}{nt}} - 1\right)$$

The following function will calculate this value for us:

```
(define (int-rate v p t)
  (let ([n 12.0])
    (* n (- (expt (/ v p) (/ 1 (* n t))) 1))))
```

So over 35 years, $1 invested in ExxonMobil stock would yield an interest rate of . . .

```
> (int-rate 18 1 35)
0.08286686131778254
```

Or around 8 percent, which is quite respectable given that ExxonMobil also pays quarterly dividends, which sweetens the pot even more.

But what about that wild stallion, Apple? Well, let's adjust our range a bit and see what develops.

```
> (plot-symbols '("^GSPC" "AAPL" "XOM") "Monthly"
                (year-range "1981" "2016")
                #t 'top-left 0 300)
```

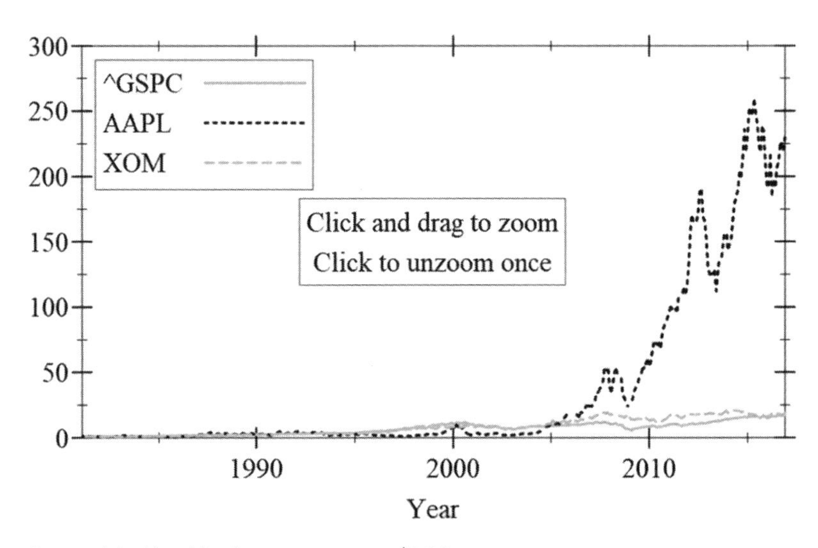

Figure 6-6: Monthly close prices up to $300

Wow, a dollar invested in 1981 (after a long dry spell) would be worth about $225 at the end of 2016. Who knew? Let's see what the effective interest rate is.

```
> (int-rate 225 1 35)
0.15574778848870174
```

Almost 16 percent—not too shabby. Is Apple a good investment? Well, any (honest) financial advisor will tell you this repeatedly: past performance is not a guarantee of future results.

Lumping Things Together

One thing besides the outrageous return on investment (ROI) of Apple stands out: its value is very erratic. Let's look at ExxonMobil and Apple during the period of time before things got really crazy in 2008 (see Figure 6-7).

```
> (plot-symbols '("AAPL" "XOM") "Daily"
                (date-range "2007-01-01" "2008-09-30")
                #t 'top-left 0.5 2.5)
```

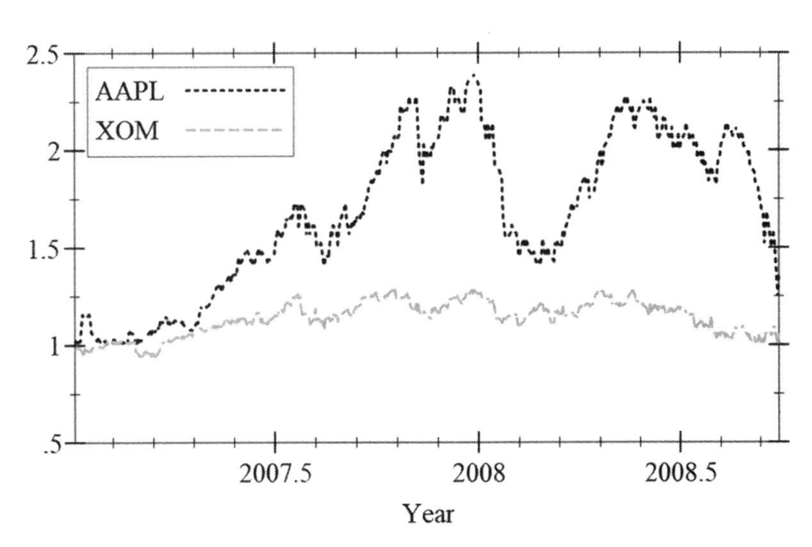

Figure 6-7: Apple and ExxonMobil before the crash

It's clear that even over this short interval of time, the value of Apple from one day to the next would not be easy to predict. If someone isn't interested in the specific value of a stock, but rather how variable it is, one way to visualize the variability is with a histogram. A *histogram* represents data by showing how often values fall into certain ranges. We call these ranges *bins*. To aid in our analysis, we'll supplement each histogram with a table that gives the range of values and counts for each bin.

To begin, we define a function that extracts a particular data field for an investment given the stock symbol, period, filter function, and the corresponding hist structure field.

```
(define (get-data symbol period filter field)
  (let ([in-port (open-input-file (data-path symbol period))])
    (read-line in-port)
    (let* ([recs
            (reverse
             (let loop([rec-list '()])
               (let ([rec (import-record in-port)])
                 (if (eof-object? rec)
                     rec-list
                     (if (filter rec)
```

```
                    (loop (cons (field rec) rec-list))
                    (loop rec-list))))))])
    (close-input-port in-port)
    recs)))
```

For example if we wanted the maximum stock cost for Microsoft in 1999 (a pretty good year for Microsoft) in the monthly data, we could get it this way:

```
> (get-data "MSFT" "Monthly"
            (year-range "1999" "1999")
            hist-close)
'(43.75 37.53125 44.8125 40.65625 40.34375 45.09375 42.90625 46.28125 45.28125
    46.28125 45.523399 58.375)
```

The following function compiles the data (a list of values) and populates bins with the proper count of values.

```
(define (categorize data avg num-bins)
  (let* ([bin (make-vector num-bins 0)]
         [bin-min (* 0.4 avg)]
         [bin-max (* 1.6 avg)]
         [bin-delta (/ (- bin-max bin-min) num-bins)])
    (define (update-bin val)
      (when (<= bin-min val bin-max)
        (let ([i (inexact->exact (floor (/ (- val bin-min) bin-delta)))])
          (vector-set! bin i (add1 (vector-ref bin i))))))
    (let loop ([val-list data])
      (unless (null? val-list)
        (update-bin (car val-list))
        (loop (cdr val-list))))
    (values bin-min bin-max
            (for/list ([i (in-range num-bins)])
              (vector i (vector-ref bin i))))))
```

This function first sets an overall range of values that are 60 percent below and 60 percent above the average value. Within this range, data values will be compiled into the bin vector. At the end of the process, the function returns the overall minimum and maximum range of the bins as well as categorized values contained in bin. Each vector in bin contains the bin index and the number of values in the bin (it has to be formatted this way to work with Racket's discrete-histogram function).

To display the data in tabular form, we define bin-table, which will display the bin index along with the range of values for the bin and the number of values in each bin.

```
(define (bin-table bins bin-min bin-max)
  (let* ([num-bins (length bins)]
         [bin-delta (/ (- bin-max bin-min) num-bins)]
         [rows
```

```
        (for/list ([i (in-range num-bins)]
                   [bin bins])
          (let ([bmin (+ bin-min (* bin-delta i))]
                [bmax (+ bin-min (* bin-delta (add1 i)))]
                [count (vector-ref bin 1)])
            (list
             (~r i #:min-width 3)
             (~r bmin #:min-width 8  #:precision (list '= 2))
             (~r bmax #:min-width 8  #:precision (list '= 2))
             (~r count #:min-width 4))))])
  (table->string (cons '("Bin" "   Min" "   Max" "Vals") rows))))
```

After we've laid the groundwork, creating a function to generate the output is quite simple.

```
(define (histogram-symbol symbol period filter [bins 11])
  (let*-values ([(data) (get-data symbol period filter hist-close)]
                [(avg) (/ (apply + data) (length data))]
                [(bin-min bin-max hist-data) (categorize data avg bins)])
    (displayln (bin-table hist-data bin-min bin-max))
    (parameterize
        ([plot-width 400]
         [plot-height 250]
         [plot-x-label #f]
         [plot-y-label "Frequency"])
      (plot (discrete-histogram hist-data)))))
```

Let's see what this tells us about Apple and ExxonMobil (see Figures 6-8 and 6-9).

```
> (histogram-symbol "AAPL" "Daily" (date-range "2007-01-01" "2008-09-30"))
+---+--------+--------+----+
|Bin|   Min  |   Max  |Vals|
+---+--------+--------+----+
|  0|    8.06|   10.25|   0|
+---+--------+--------+----+
|  1|   10.25|   12.45|  30|
+---+--------+--------+----+
|  2|   12.45|   14.65|  55|
+---+--------+--------+----+
|  3|   14.65|   16.85|  20|
+---+--------+--------+----+
|  4|   16.85|   19.04|  87|
+---+--------+--------+----+
|  5|   19.04|   21.24|  54|
+---+--------+--------+----+
|  6|   21.24|   23.44|  46|
+---+--------+--------+----+
|  7|   23.44|   25.64|  81|
```

```
+---+--------+--------+----+
|  8|   25.64|   27.83|  61|
+---+--------+--------+----+
|  9|   27.83|   30.03|   6|
+---+--------+--------+----+
| 10|   30.03|   32.23|   0|
+---+--------+--------+----+
```

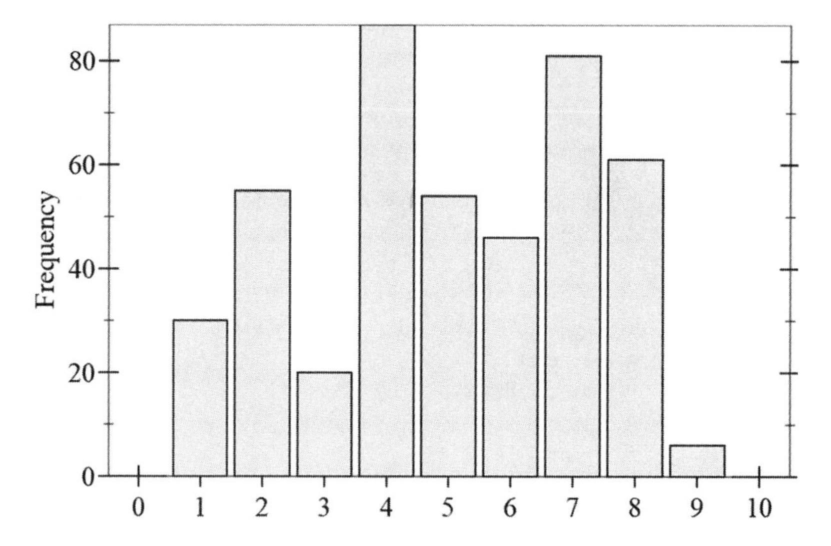

Figure 6-8: Apple histogram

```
> (histogram-symbol "XOM" "Daily" (date-range "2007-01-01" "2008-09-30"))

+---+--------+--------+----+
|Bin|   Min  |   Max  |Vals|
+---+--------+--------+----+
|  0|   33.65|   42.83|   0|
+---+--------+--------+----+
|  1|   42.83|   52.00|   0|
+---+--------+--------+----+
|  2|   52.00|   61.18|   0|
+---+--------+--------+----+
|  3|   61.18|   70.36|   4|
+---+--------+--------+----+
|  4|   70.36|   79.54| 106|
+---+--------+--------+----+
|  5|   79.54|   88.71| 210|
+---+--------+--------+----+
|  6|   88.71|   97.89| 120|
+---+--------+--------+----+
|  7|   97.89|  107.07|   0|
```

```
+---+--------+--------+----+
| 8| 107.07| 116.25|   0|
+---+--------+--------+----+
| 9| 116.25| 125.42|   0|
+---+--------+--------+----+
| 10| 125.42| 134.60|   0|
+---+--------+--------+----+
```

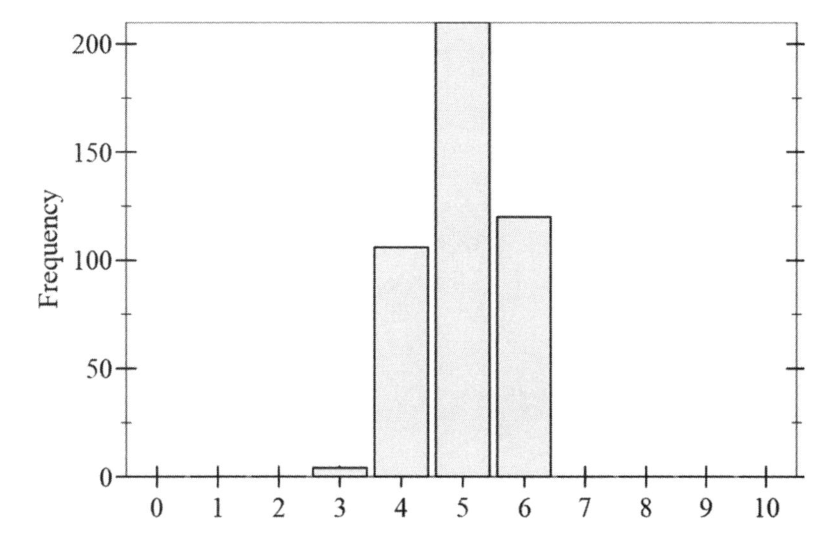

Figure 6-9: ExxonMobil histogram

It's clear from the histograms that values for Apple are dispersed over a wider range of values than the data for ExxonMobil for the same time period. This higher volatility is the price Apple investors pay for the possibility of getting larger returns.

A Bit of Statistics

We demonstrated in the previous section that the old truism that a picture is worth a thousand words makes a bit of sense, at least in terms of analyzing investment data. But it's also true that a single number is worth, well, at least one picture. So far our analysis has primarily been qualitative, where we used a number of techniques to visualize our data. We now turn to a couple of standard statistical tools that are widely used for quantitative analysis.

Standard Deviation

In the last section, we looked at using histograms to get some idea of how widely dispersed stock values were over some period of time. This type of information can be summed up in a single number called the *standard deviation*. For a given set of numbers, the standard deviation indicates how much individual numbers deviate from the overall average of the set. You might think of it as the average amount of deviation. Standard deviation is defined by the following formula:

$$\sigma = \sqrt{\frac{1}{n}\sum_{i=1}^{n}(x_i - \mu)^2}$$

In this equation, n is the number of values, the Greek letter mu (μ) represents the mean or average of all the data values, and the x_i represents the individual numbers.

Closely related to standard deviation is the statistical concept of *variance*, which is simply the square of the standard deviation:

$$\sigma_x^2 = \frac{1}{n}\sum_{i=1}^{n}(x_i - \mu)^2$$

We shall see a bit later that variance is useful in regression analysis where we attempt to determine how data is trending.

We wrap the standard deviation formula in a Racket function as follows:

```
(define (std-deviation nums)
  (let* ([n (length nums)]
         [mu (/ (apply + nums) n)]
         [sqr-diff (map (λ (x) (sqr (- x mu))) nums)])
    (sqrt (/ (apply + sqr-diff) n))))
```

We can now compute numerical values with which we can analyze the deviation of different assets. Let's take a look at the data we generated histograms for.

```
> (define apple (get-data "AAPL" "Daily" (date-range "2007-01-01" "2008-09-30"
    ) hist-close))
> (define xom (get-data "XOM" "Daily" (date-range "2007-01-01" "2008-09-30")
    hist-close))

> (std-deviation apple)
4.811932439819516

> (std-deviation xom)
6.399636764602135
```

This *seems* to indicate that Apple actually had less deviation than ExxonMobil. This is where the proper interpretation of data is crucial. The histogram data showed us the spread of data within ±60% of the *average* value. To make sense of the deviation data, let's get the average value of these stocks.

```
> (define apple-avg (/ (apply + apple) (length apple)))
> apple-avg
20.143350647727257
```

```
> (define xom-avg (/ (apply + xom) (length xom)))
> xom-avg
84.12513634318191

> (/ (std-deviation apple) apple-avg)
0.23888441024395504

> (/ (std-deviation xom) xom-avg)
0.07607282487478317
```

From this we can now see that Apple typically deviated almost 24 percent from its average price during that period whereas ExxonMobil only deviated about 7.5 percent.

While here we're strictly looking at the standard deviation of the close price, this is *not* how deviation is normally evaluated in a financial sense. What is of most interest in that regard is deviation of *returns*. A stock that has a steady 10 percent annual return would clearly have some price deviation, but it would have almost no deviation based on returns. Another consideration is that change in stock price alone is not necessarily an indication of returns since dividends (for those stocks that pay dividends) come into play as well.

Regression

In our analysis of various financial assets, we've mentioned the adage that past performance is not a guarantee of future results, which is true, but past performance might suggest future results. Given a disparate set of data points, it's often of interest to determine whether or not they suggest a trend. The statistical tool called *regression analysis* attempts to make just such a determination. Regression analysis fits a straight line to a set of data points (since we're only fitting our data to a straight line, this is technically called *linear regression*) where x is called the *independent predictor* or *predictor variable*, and y is called the *dependent response*. The desired outcome is this *regression* or *prediction* line:

$$y = a + bx$$

The idea is that given this line and some x-value, we can compute an estimated value for y. The parameters a and b are defined in such a way as to minimize the total distance of the y data values from the line. Specifically, if (x_i, y_i) are the actual data points, we let $\hat{y}_i = a + bx_i$ (this is the estimated value of y at x_i), in which case regression analysis seeks to minimize the *sum of squares errors* given by the following:

$$\text{SSE} = \sum_{i=1}^{n} \left(y_i - \hat{y}_i\right)^2$$

If $\bar{x}$ is the mean, or average, of all the x values and $\bar{y}$ is the mean of all the y values, it can be shown that the a and b parameters of the regression line are given by the following:

$$b = \frac{\sum_{i=1}^{n} (x_i - \bar{x})(y_i - \bar{y})}{\sum_{i=1}^{n} (x_i - \bar{x})^2} \tag{6.1}$$

$$a = \bar{y} - b\bar{x} \tag{6.2}$$

The formidable-looking Equation (6.1) is actually the ratio of two simpler formulas: the *covariance* of x and y and the variance of x. We've already seen that the variance of x is given by the following:

$$\sigma_x^2 = \frac{1}{n}\sum_{i=1}^{n} (x_i - \bar{x})^2$$

The covariance of x and y is defined by the following:

$$\mathrm{Cov}(x, y) = \frac{1}{n}\sum_{i=1}^{n} (x_i - \bar{x})(y_i - \bar{y})$$

Covariance is a measure of the joint variability of two random variables (in our case the x_i and y_i). We make use of the following slightly altered form of these last two equations, along with a couple of others, to develop a method to determine how well our regression line actually fits the data.

$$\mathrm{SS}_{xx} = \sum_{i=1}^{n} (x_i - \bar{x})^2$$

$$\mathrm{SS}_{xy} = \sum_{i=1}^{n} (x_i - \bar{x})(y_i - \bar{y})$$

$$\mathrm{SS}_{yy} = \sum_{i=1}^{n} (y_i - \bar{y})^2$$

$$\mathrm{SSR} = \sum_{i=1}^{n} (\hat{y}_i - \bar{y})^2$$

We see that SS_{xx} is just a slightly tweaked version of σ_x^2 and SS_{xy} is likewise a tweaked version of $\mathrm{Cov}(x, y)$. The last equation, sum of squares regression, represents the sum of the squares of the distance of the estimated $\hat{y}$'s from the mean of the y-values to the regression line (the line that minimizes this distance will give us the best fit to the data). We saw in Equation (6.1) that the slope of the regression line is given by the following:

$$b = \frac{\mathrm{SS}_{xy}}{\mathrm{SS}_{xx}} \tag{6.3}$$

But the following can also be shown:

$$SS_{yy} = SSR + SSE$$

$$\frac{SSR}{SS_{yy}} = 1 - \frac{SSE}{SS_{yy}} = R^2$$

The last equation is called the *squared correlation* or *coefficient of determination*. This number can vary between 0 and 1. A value of 1 indicates the data points perfectly fit the regression line, and a value of 0 indicates there's no correlation.

The regression line parameters can be computed with the following Racket function:

```
(define (regression-params data)
  (define (x v) (vector-ref v 0))
  (define (y v) (vector-ref v 1))
  (let* ([num (length data)]
         [totx (apply + (map x data))]
         [toty (apply + (map y data))]
         [avgx (/ totx num)]
         [avgy (/ toty num)]
         [ss-xy (apply + (map (λ (v) (* (- (x v) avgx) (- (y v) avgy))) data))]
         [ss-xx (apply + (map (λ (v) (sqr (- (x v) avgx))) data))]
         [b (/ ss-xy ss-xx)]
         [a (- avgy (* b avgx))])
    (values a b)))
```

This is a straightforward adaption of Equations (6.1) and (6.2). Let's see what this tells us about Bank of America when it was in its downward spiral during the 2008 financial crisis.

```
> (define bac (get-coords "BAC" "Monthly"
                          (date-range "2007-07-01" "2009-02-01")
                          #f))

> (regression-params bac)
54422.310899480566
-27.082265190974677
```

The second value is the slope of the regression line, which indicates that during that period, it was losing an average of $27 per year (over half its value in July of 2007). Ouch.

We now define a plot routine that accepts a single asset symbol, but that plots the data points instead of lines and includes the corresponding regression line.

```
(define (plot-regression symbol period filter
                         [anchor 'top-left])
  (let* ([coords (get-coords symbol period filter #f)]
```

```
        [plot-data
         (let ([color (hash-ref symbol-color symbol)])
           (points coords #:label symbol #:color color))])
    (let-values ([(a b) (regression-params coords)])
      (parameterize
          ([plot-width 400]
           [plot-height 250]
           [plot-x-label "Year"]
           [plot-y-label #f]
           [plot-legend-anchor anchor])
        (plot (list
                plot-data
                (function (λ (x) (+ (* b x) a))
                          #:color "black" #:label "Regr")))))))
```

We can see Bank of America's distress in graphic detail in Figure 6-10.

```
> (plot-regression "BAC" "Monthly"
                   (date-range "2007-07-01" "2009-02-01")
                   'bottom-left)
```

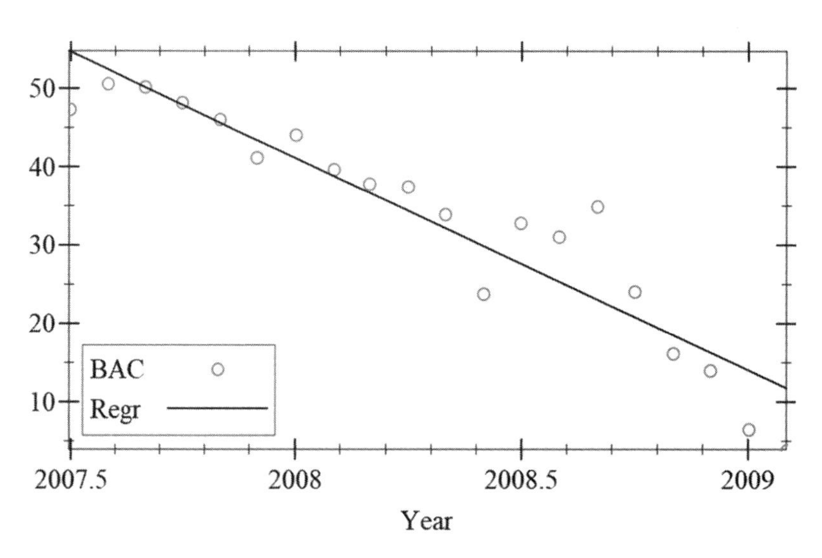

Figure 6-10: Regression line plotted on Bank of America data

To determine how well the regression line fits the data, we define a correlation function:

```
(define (correlation data)
  (define (x v) (vector-ref v 0))
  (define (y v) (vector-ref v 1))
  (let* ([num (length data)]
         [totx (apply + (map x data))]
         [toty (apply + (map y data))]
```

```
         [avgx (/ totx num)]
         [avgy (/ toty num)]
         [ss-xx (apply + (map (λ (v) (sqr (- (x v) avgx))) data))]
         [ss-yy (apply + (map (λ (v) (sqr (- (y v) avgy))) data))]
         [ss-xy (apply + (map (λ (v) (* (- (x v) avgx) (- (y v) avgy))) data))
  ]
         [b (/ ss-xy ss-xx)]
         [a (- avgy (* b avgx))]
         [ssr (apply + (map (λ (v) (sqr (- (+ (* b (x v)) a) avgy))) data))])
   (/ ssr ss-yy)))
```

This is also just a direct implementation of the definition of R^2 given above. With this we can test the fit of the Bank of America least squares line to the data.

```
> (define bac (get-coords "BAC" "Monthly"
                          (date-range "2007-07-01" "2009-02-01")
                          #f))
> (correlation bac)
0.8799353920116734
```

This indicates a fairly good fit. But in many cases the data doesn't provide a good fit with a straight line. For instance if we include the start of the recovery phase, we wind up with a plot such as the following in Figure 6-11.

```
> (plot-regression "BAC" "Monthly"
                   (year-range "2008" "2009")
                   'bottom-left)
```

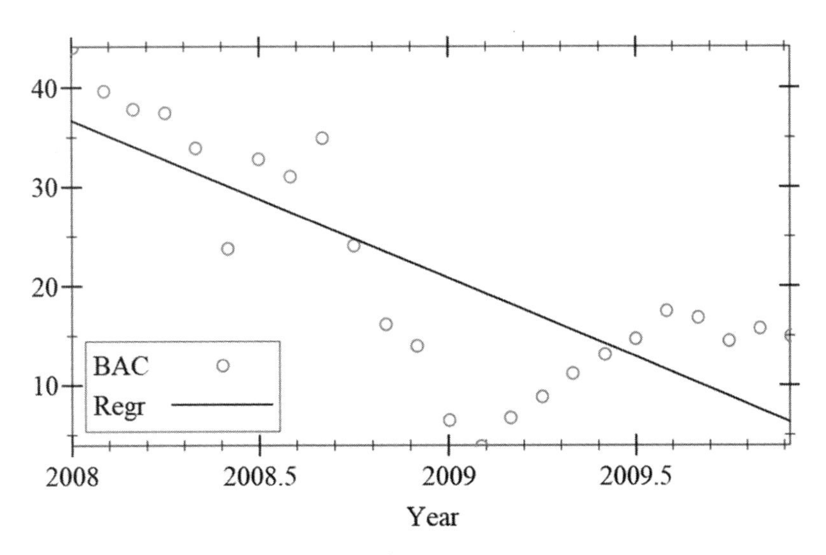

Figure 6-11: Poorly fitting regression line

And this indicates some degree of correlation, but not as good as what we had before:

```
> (define bac (get-coords "BAC" "Monthly"
                          (year-range "2008" "2009")
                          #f))

> (correlation bac)
0.6064135484684874
```

Summary

In this chapter, we looked at various ways of accessing and analyzing data using Racket and DrRacket. We began by introducing the mechanics of how to import and export data to and from Racket ports. Once we had mastered this bit of technology, we leveraged it to look at securities in the form of raw historical stock market values. We then took a slight detour and explored binary search using random file access. Having defined a mechanism to access and parse our stock market data, we then looked at qualitative methods to analyze our data using various visualization techniques. Finally, we bit the bullet and introduced a bit of mathematics, which allowed us to do some statistical quantitative analysis.

Next up, we see how we can use some sophisticated search algorithms to solve some classical problems in recreational mathematics.

7

SEARCHING FOR ANSWERS

For all the problems we have encountered so far, there was a direct method to compute the solution. But this is not always the case. For many problems, we have to search for the solution using some type of algorithm, such as when solving a Sudoku puzzle or the *n*-queens problem. In these cases, the process involves trying a series of steps until either we find the solution or we have to back up to a previous step to try an alternative route. In this chapter we'll explore a number of algorithms that allow us to efficiently select a path that leads to a solution. Such an approach is known as a *heuristic*. In general, a heuristic isn't guaranteed to find a solution, but the algorithms we explore here (thankfully) are.

Graph Theory

It's often the case that a problem we're trying to solve can be modeled with a *graph*. Intuitively, a graph is just a set of points (or nodes) and connecting lines, as illustrated in Figure 7-1. Each node represents some state of the problem-solving process, and the lines extending from one node to other nodes represent possible alternative steps. We'll first give some basic graph definitions as background before delving into the actual problem-solving algorithms.

The Basics

Formally, a graph is a finite set V of *vertices* (or nodes) and a set E of *edges* joining different pairs of distinct vertices (see Figure 7-1).

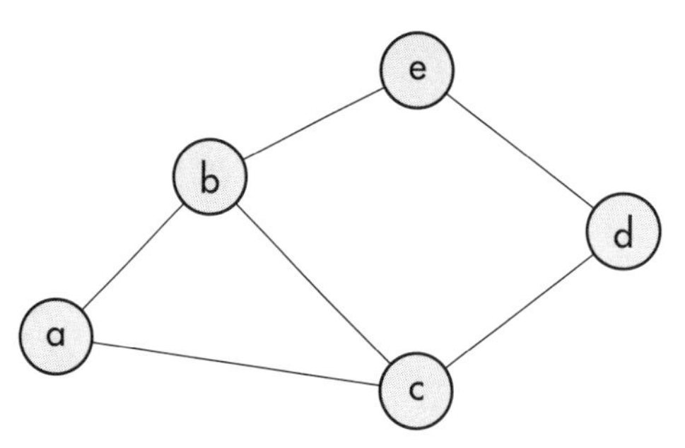

Figure 7-1: Graph

In Figure 7-1 above, $V = \{a, b, c, d, e\}$ are the vertices and $E = \{(a, b), (a, c), (b, c), (c, d), (b, e), (e, d)\}$ are the edges.

A sequence of graph vertices $(v_1, v_2, \ldots, v_n)$, such that there's an edge connecting v_i and v_{i+1}, is called a *walk*. If all the vertices are distinct, a walk is called a *path*. A walk where all the vertices are distinct except that $v_1 = v_n$ is called a *cycle* or *circuit*. In Figure 7-1, the sequence (a, b, c, b) is a walk, the sequence (a, b, c, d) is a path, and the sequence (a, b, c, a) is a cycle.

A graph with a path from each vertex to every other vertex is said to be *connected*. A connected graph without any cycles is called a *tree*. In a tree, any path is assumed to flow from upper nodes to lower nodes. Such a structure (where there are no cycles and there is only one way to get from one node to another) is known as a *directed acyclic graph (DAG)*. It's possible to convert the graph above to a tree by removing some of its edges, as shown in Figure 7-2.

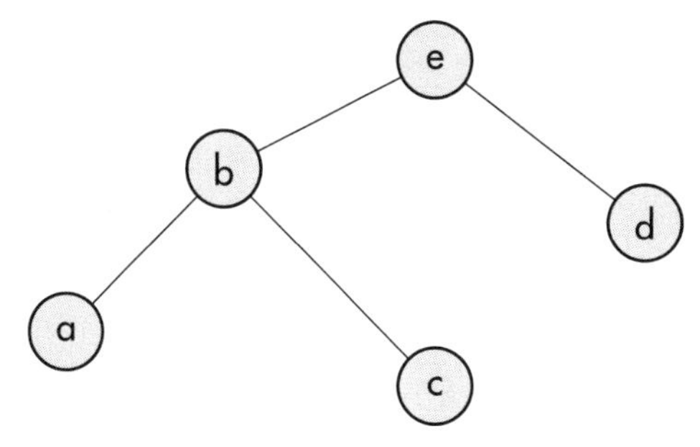

Figure 7-2: Tree

If nodes *x* and *y* are connected in such a way that it's possible to go from *x* to *y*, then *y* is said to be a *child node* of *x*. Nodes without child nodes (such as *a*, *c*, and *d*) are known as *terminal* (or *leaf*) nodes. Problems that have solutions modeled by a tree structure lend themselves to simpler search strategies since a tree doesn't have circuits. Searching a graph with circuits requires keeping track of nodes already visited so that the same nodes aren't re-explored.

It's possible to label each edge of the graph with a numerical value called a *weight*, as shown in Figure 7-3. This type of graph is called a *weighted graph*.

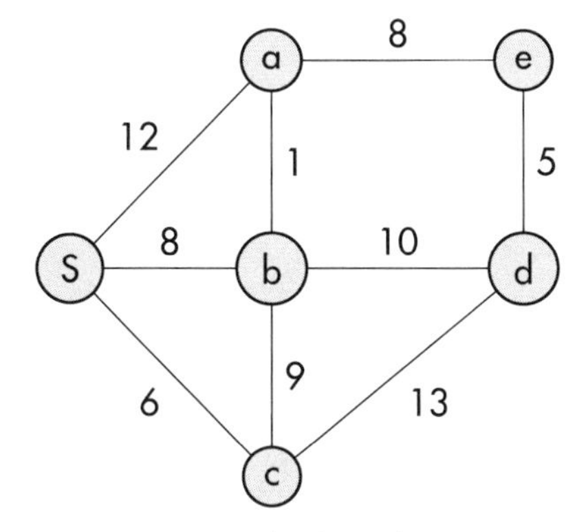

Figure 7-3: Weighted graph

If *e* is an edge, the weight of the edge is designated by *w*(*e*). Weights can be used to represent any number of measurements such as time, cost, or distance, which may affect the choice of edge when searching a graph.

A number of interesting questions arise when exploring the properties of graphs. One such question is this: "Given any two nodes, what's the shortest path between them?" Another is the famous traveling salesman problem: "Given a list of cities and the distances between them, what's the shortest possible route that visits each city exactly once and returns to the original city?" This last question, where each node is visited exactly once and returns to the original node, involves what is called a *Hamiltonian circuit*.

Graph Search

There are two broad categories of strategies for searching graphs: *breadth-first search (BFS)* and *depth-first search (DFS)*. To illustrate these concepts, we'll use the tree in Figure 7-2.

Breadth-First Search

Breadth-first search involves searching a graph by fully exploring each level (or depth) before moving on to the next level. In the tree diagram (shown in Figure 7-2), the *e* (root) node is on the first level, nodes *b* and *d* are on the next level, and nodes *a* and *c* are on the third level. This typically involves using a queue to stage the nodes to be examined. The process begins by pushing the root node onto the queue, as shown in Figure 7-4:

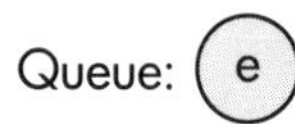

Figure 7-4: A queue containing the root node

We then pop the first node in the queue (*e*) and test it to see if it's a goal node; if not, we push its child nodes onto the queue, as shown in Figure 7-5:

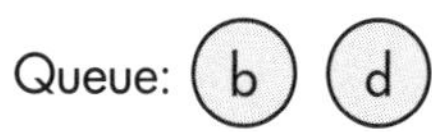

Figure 7-5: The queue after node e *was explored*

Again we pop the first node in the queue (this time *b*) and test it to see if it's a goal node; if not, we push its child nodes onto the queue, as shown in Figure 7-6:

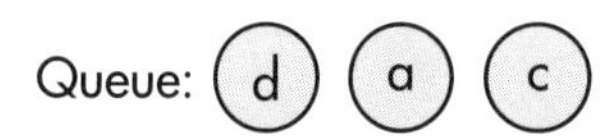

Figure 7-6: The queue after node b *was explored*

We continue in this fashion until a goal node has been found, or the queue is empty, in which case there's no solution.

Depth-First Search

Depth-first search works by continuing to walk down a branch of the tree until a goal node is found or a terminal node is reached. For example, starting at the root node of the tree, nodes *e*, *b*, and *a* would be examined in order. If none of those node are goal nodes, we back up to node *b* and examine its next child node, *c*. If *c* is also not a goal node, we back all the way up to *e* and examine its next child node, *d*. The *n*-queens problem in the next section provides a simple example of using depth-first search.

The *N*-Queens Problem

The *n*-queens problem is a classic problem often used to illustrate depth-first search. The problem goes like this: position *n* queens on an *n*-by-*n* chessboard such that no queen is attacked by any other queen. In case you aren't familiar with chess, a queen can attack any square on the same row, column, or diagonal that the queen lies on, as illustrated in Figure 7-7.

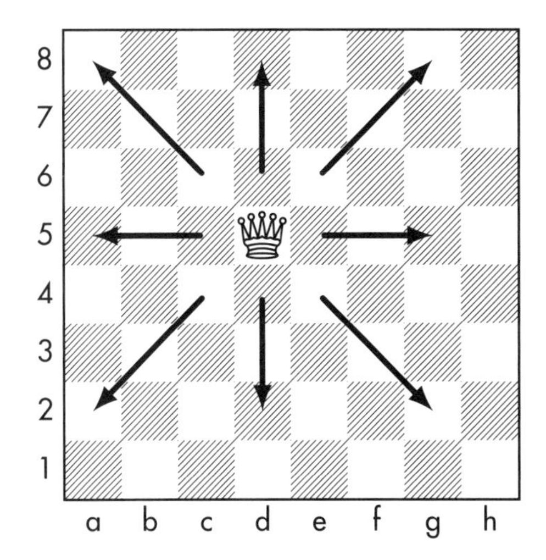

Figure 7-7: A queen's possible moves

The smallest value of *n* for which a solution exists is 4. The two possible solutions are shown in Figure 7-8.

Figure 7-8: Solutions to the 4-queens problem

One reason for the popularity of this problem is that the search graph is a tree, meaning that with a depth-first search, there's no possibility that a state previously seen will be reached again (that is, once a queen is placed, it's not possible to get to a state with fewer queens in subsequent steps). This avoids the annoying need to keep track of previous states to ensure they aren't explored again.

A simple approach to this problem is to go column by column, testing each square in a column and continuing until a solution has been reached (backtracking as required). For example, if we begin with Figure 7-9, we can't place a queen at b1 or b2, because it would be attacked by the queen at a1.

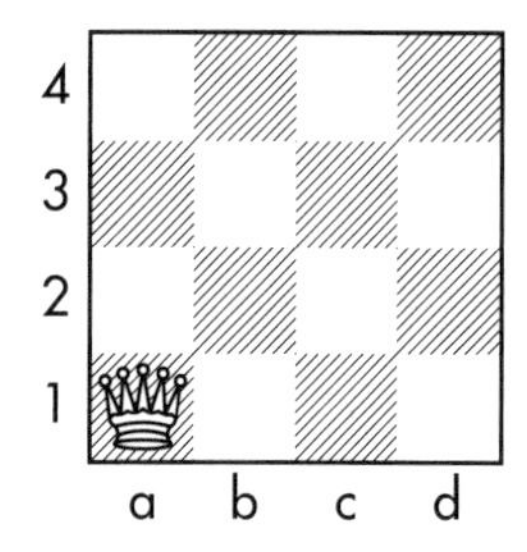

Figure 7-9: First queen at a1

The next available square that's not attacked is b3, resulting in Figure 7-10:

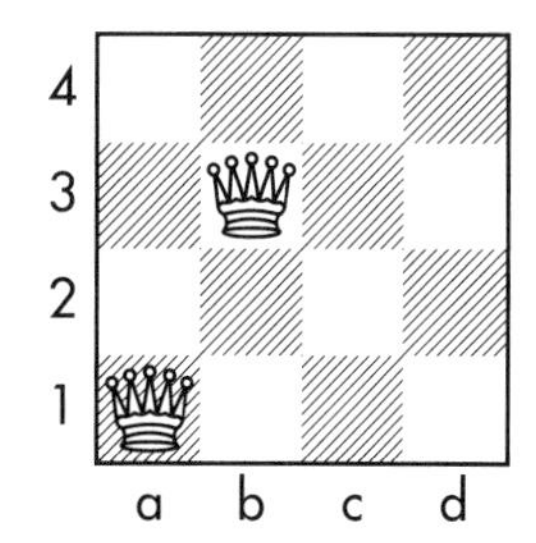

Figure 7-10: Second queen at b3

But now when we get to column c, we're stuck since every square in that column is attacked by one of the other queens. So we backtrack and move the queen in column b to b4 in Figure 7-11:

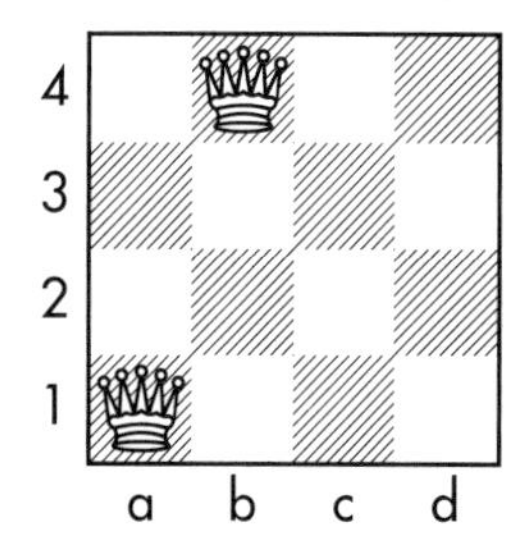

Figure 7-11: Second queen at b4

So now we can place a queen on c2 in Figure 7-12:

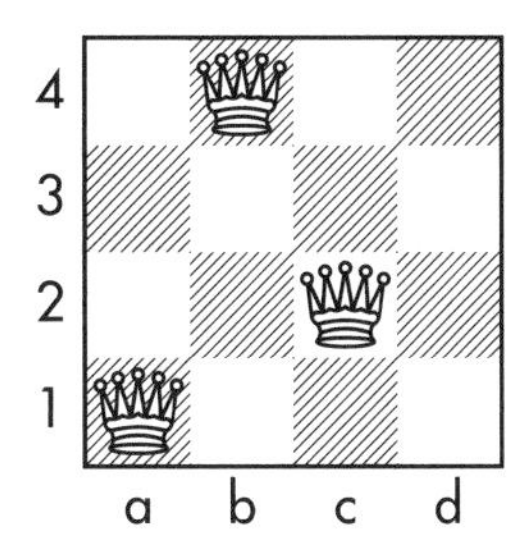

Figure 7-12: Third queen at c2

Alas, now there's no spot for a queen on column d. So we backtrack all the way to column a and start over in Figure 7-13.

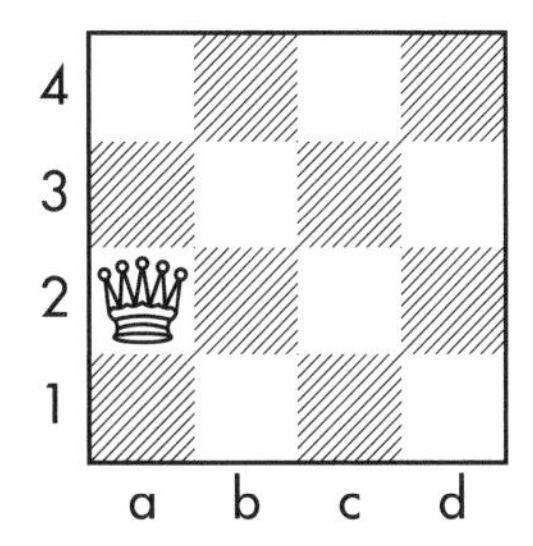

Figure 7-13: Backtrack to first queen at a2

The process continues in this manner until a solution is found.

A Racket Solution

We define the chessboard as an *n*-by-*n* array constructed from a mutable vector with *n* elements, each of which is also an *n*-element vector, where each element is either a 1 or a 0 (0 means the square is unoccupied; 1 means the square has a queen):

```
(define (make-chessboard n)
  (let loop ([v n] [l '()])
    (if (zero? v)
        (list->vector l)
        (loop (sub1 v) (cons (make-vector n 0) l)))))
```

To allow accessing elements of the chessboard `cb` by a row (`r`) and column (`c`) number, we define the following accessor forms, where `v` is the value being set or retrieved.

```
(define (cb-set! cb r c v)
  (vector-set! (vector-ref cb c) r v))

(define (cb-ref cb r c)
  (vector-ref (vector-ref cb c) r))
```

Since we're using a mutable data structure for the chessboard, we'll need a mechanism to copy the board whenever a solution is found, to preserve the state of the board.

```
(define (cb-copy cb)
  (for/vector ([v cb]) (vector-copy v)))
```

We'll of course need to be able to see the solutions, so we provide a print procedure:

```
(define (cb-print cb)
  (let ([n (vector-length cb)])
    (for* ([r n]
           [c n])
      (when (zero? c) (newline))
      (let ([v (cb-ref cb r c)])
        (if (zero? v)
            (display " .")
            (display " Q")
            ))))
  (newline))
```

The actual code to solve the problem, `dfs`, is a straightforward depth-first search. As solutions are found, they're compiled into a list called `sols`, which is the return value of the function. In the code below, recall that in the `let loop` form, we're employing a named `let` (which we described in Chapter 3) where we're defining a function (`loop`) that we'll call recursively.

```
(define (dfs n)
  (let ([sols '()]
        [cb (make-chessboard n)])
    (let loop([r 0][c 0])
      (when (< c n)
     ❶ (let ([valid (not (attacked cb r c))])
          (when valid
         ❷ (cb-set! cb r c 1)
         ❸ (if (= c (sub1 n))
                (let ([copy (cb-copy cb)])
               ❹ (set! sols (cons copy sols)))
              ❺ (loop 0 (add1 c)))
         ❻ (cb-set! cb r c 0))
       ❼ (when (< (add1 r) n) (loop (add1 r) c)))))
 ❽ sols))
```

The code first tests each position to see whether the current cell is being attacked by any of the queens that have already been placed on the board ❶ (the code for `attacked` will be described shortly); if not, then that cell is marked as `valid`, and a queen (the number 1) is placed on that square ❷. Next we test whether the current square is in the final column of the board ❸; if it is, we've found a solution, so a copy of the board is placed in `sols` ❹. If we're not on the last column, we then nest down to the next level (that is, the next column) ❺. Finally, the valid square is cleared ❻ so that additional rows in the column can be tested ❼. Once all the solutions have been found, they're returned ❽.

Where the DFS backtracking occurs in this process is a bit subtle. Suppose we're at a position that is under attack by the previously placed queens, so `valid` ❶ is false and execution falls through ❼. Now suppose we're also on the last row. In that case, the test fails ❼, so no further looping occurs and the recursive call returns. Either there are no following statements, in which case the entire loops exits, or there there additional statements to execute after returning from the recursive call. This can only occur where the current position is cleared and we're at a previous location ❻. This is the backtrack point. Execution then resumes at the last `when` statement ❼.

The following function tests whether a square is under attack by any of the previously placed queens. It only checks the columns prior to the current column since the other columns of the chessboard haven't yet been populated.

```
(define (attacked cb r c)
  (let ([n (vector-length cb)])
    (let loop ([ac (sub1 c)])
      (if (< ac 0) #f
          (let ([r1 (+ r (- c ac))]
                [r2 (+ r (- ac c))])
            (if (or (= 1 (cb-ref cb r ac))
                    (and (< r1 n) (= 1 (cb-ref cb r1 ac)))
                    (and (>= r2 0) (= 1 (cb-ref cb r2 ac))))
                #t
                (loop (sub1 ac))))))))
```

To output the solutions, we define a simple routine to iterate through and print each solution returned by `dfs`.

```
(define (solve n)
  (for ([cb (dfs n)]) (cb-print cb)))
```

Here are a couple of test runs.

```
> (solve 4)

 . Q . .
 . . . Q
 Q . . .
```

```
. . Q .

. . Q .
Q . . .
. . . Q
. Q . .

> (solve 5)

. . Q . .  . . . Q .  . . . . Q  . Q . . .  . . . . Q
. . . . Q  . Q . . .  . Q . . .  . . . . Q  . . Q . .
. Q . . .  . . . . Q  . . . Q .  . . Q . .  Q . . . .
. . . Q .  . . Q . .  Q . . . .  Q . . . .  . . . Q .
Q . . . .  Q . . . .  . . Q . .  . . . Q .  . Q . . .

. Q . . .  . . . Q .  . . Q . .  Q . . . .  Q . . . .
. . . Q .  Q . . . .  Q . . . .  . . Q . .  . . . Q .
Q . . . .  . . Q . .  . . . Q .  . . . . Q  . Q . . .
. . Q . .  . . . . Q  . Q . . .  . Q . . .  . . . . Q
. . . . Q  . Q . . .  . . . . Q  . . . Q .  . . Q . .

> (solve 8)

. . Q . . . . .
. . . . . Q . .
. . . Q . . . .
. Q . . . . . .
. . . . . . . Q
. . . . Q . . .
. . . . . . Q .
Q . . . . . . .

<intermediate solutions omitted>

Q . . . . . . .
. . . . . . Q .
. . . . Q . . .
. . . . . . . Q
. Q . . . . . .
. . . Q . . . .
. . . . . Q . .
. . Q . . . . .
```

Dijkstra's Shortest Path Algorithm

Given a graph with a node designated as the start node, Edsger Dijkstra's algorithm finds the shortest path to any other node. The algorithm works by first assigning all the nodes (except the start node, which has distance zero) an infinite distance. As the algorithm progresses, the node distances are refined until their true distance can be determined.

We'll use the weighted graph introduced in Figure 7-3 earlier to illustrate Dijkstra's algorithm (where *S* is the starting node). The algorithm we describe will employ something called a priority queue. A *priority queue* is similar to a regular queue, but in a priority queue, each item has an associated value, called its priority, that controls its order in the queue. Instead of following a first-in, first-out sequence, items with a higher priority are ordered ahead of other items. Since we're interested in finding the shortest path, shorter distances will be given a higher priority than longer ones.

The following diagram in Figure 7-14 illustrates the starting conditions of the algorithm.

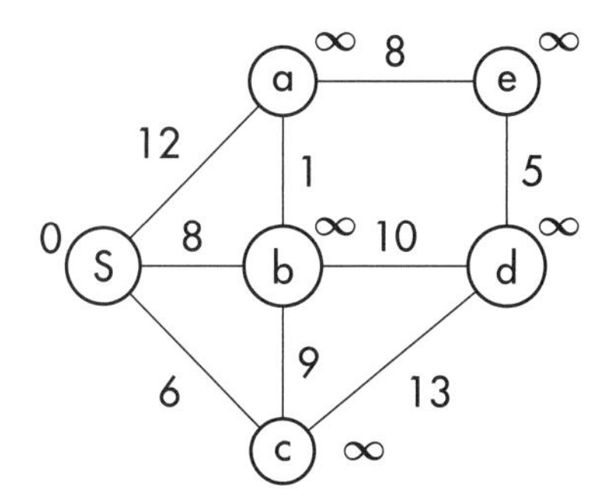

Queue: S^0, a^∞, b^∞, c^∞, d^∞, e^∞

Figure 7-14: Starting conditions for finding the shortest paths from S

Node distances from the start node are given just outside the node circle. Nodes that haven't been visited are assigned a tentative distance value of infinity (except for the start node, which has a value of zero). The queue shows the nodes with distance values indicated by exponents.

The first step is to pop the first node in the queue (which will always have a known distance) and color it with a light background as shown here in Figure 7-15. Set this node as the current node, *u* (in this case $u = S$ with a distance value of zero).

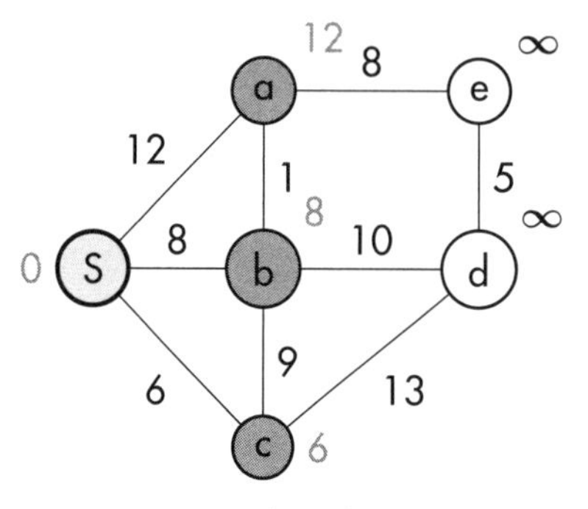

Figure 7-15: Step 1 of Dijkstra's algorithm

The neighbors of u are given in a darker color. We then perform the following tentative distance calculation, t, for each neighbor (designated v) of u that's still in the queue, where $d(u)$ is the known distance from the start node to u, and $l(u, v)$ is the distance value of the edge from u to v:

$$t = d(u) + l(u, v) \tag{7.1}$$

If t is less than the prior distance value (initially ∞), the queue is updated with the new node distance.

With the queue updated, we repeat the process, this time popping c off the queue, making it the current node (in other words, $u = c$), and updating the queue and neighbor distances as before. The state of the graph is then as follows in Figure 7-16:

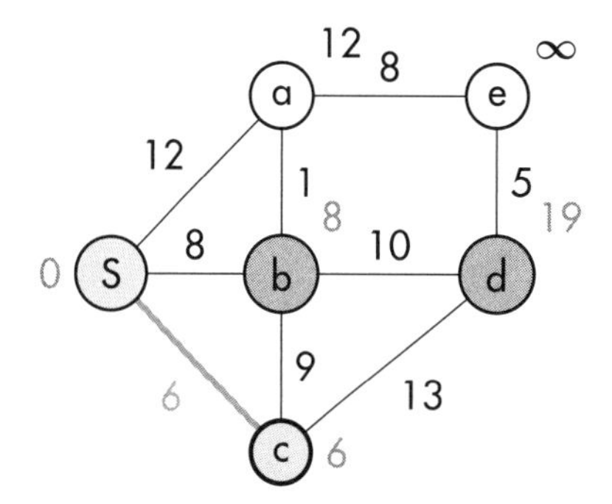

Figure 7-16: Step 2 of Dijkstra's algorithm

We show the path from S to c in a thicker gray line to indicate a known shortest path in Figure 7-16. The sequence of diagrams in Figure 7-17 illustrates the remainder of the process. Notice that in Figure 7-17a the original distance of node a has been updated from 12 to 9 based on the path now being from S through b to a. The thick lines in 7-17d, the final graph, form a tree structure reflecting all the shortest paths originating from node S to the remaining nodes.

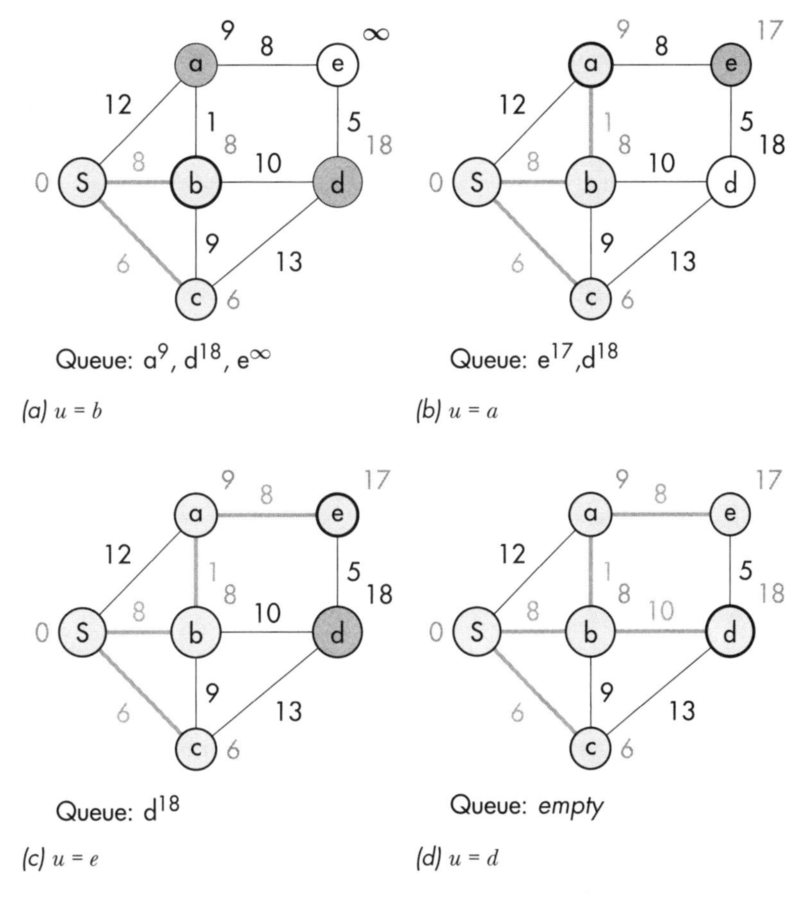

Figure 7-17: The rest of Dijkstra's algorithm

We're usually interested in how efficiently an algorithm performs. This is generally specified by a *complexity* value. There are a number of ways this can be done, but a popular formulation is called *Big O notation* (the O stands for "order of"). This notation aims to give a gross approximation of how efficiently an algorithm performs (in terms of running time or memory usage) based on the size of its inputs. Dijkstra's algorithm has a running time complexity of $O(N^2)$, where N is the number of nodes in the graph. This means the running time increases as the square of the number of inputs. In other words, if we double the number of nodes, the algorithm will take about four times as long to run. This is taken to be an upper-bound or worst-case scenario, and depending on the nature of the graph, the runtime could be less.

The Priority Queue

As we've seen in the analysis above, a priority queue plays a key role in Dijkstra's algorithm. Priority queues can be implemented in a number of ways, but one popular approach is to use something called a binary heap. A *binary heap* is a binary tree structure (meaning each node has a maximum of two children) where each node's value is greater than or equal to its child nodes. This type of heap is called a *max-heap*. It's also possible for each parent node to be less than or equal to its child nodes. This type of heap is called a *min-heap*. An example of such a heap is shown in Figure 7-18. The top or root node is always the first to be removed since it's considered to have the highest priority. After nodes are added to or removed from the heap, the remaining nodes are rearranged to maintain the proper priority order. While it's not terribly difficult to build a binary heap object, Racket already has one available in the *data/heap* library.

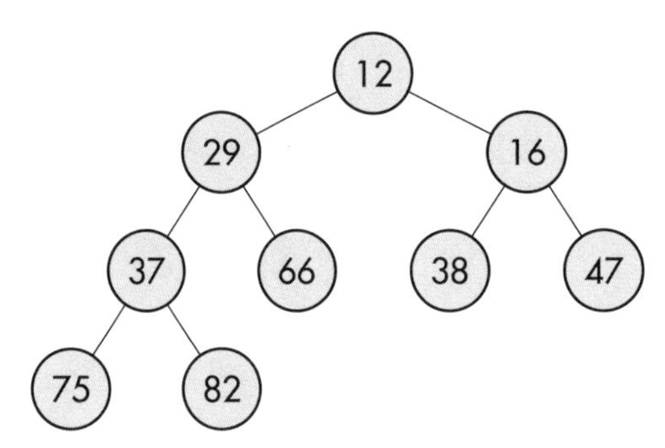

Figure 7-18: Min-heap

Our heap entries won't be just numbers: we'll need to track the node and its current distance value (which determines its priority). So each heap entry will consist of a pair where the first element is the node and the second element is the current distance. When a Racket heap is constructed, it must be supplied with a function that will do the proper comparison when given two node entries. We accomplish this with the following code. The `comp` function only compares the second element of each pair since that's what determines the priority.

```
#lang racket
(require data/heap)

(define (comp n1 n2)
  (let ([d1 (cdr n1)]
        [d2 (cdr n2)])
    (<= d1 d2)))

(define queue (make-heap comp))
```

To save a bit of typing, we create a few simple helper functions.

```
(define (enqueue n) (heap-add! queue n))

(define (dequeue)
  (let ([n (heap-min queue)])
    (heap-remove-min! queue)
    n))
```

```
(define (update-priority s p)
  (let ([q (for/first ([x (in-heap queue)] #:when (equal? s (car x))) x)])
    (heap-remove! queue q)
    (enqueue (cons s p))))

(define (peek-queue) (heap-min queue))

(define (queue->list) (for/list ([n (in-heap queue)]) n))

(define (in-queue? s)
  (for/or ([x (in-heap queue)]) (equal? (car x) s)))
```

The `update-priority` procedure takes a symbol and a new priority to update the queue. It does this by removing (dequeuing) the old value and adding (enqueuing) a new value. The `heap-remove!` function performs very efficiently, but it needs the exact value (pair with symbol and priority) to work. Unfortunately, without knowing the priority, we have to resort to a linear search through the entire queue to find the symbol via the `in-heap` sequence. This can be optimized by storing (in another data structure like a hash table) the symbol and current priority. We leave it to the reader to perform this added step if desired.

Here are some examples of the priority queue in action.

```
> (enqueue '(a . 12))
> (enqueue '(b . 8))
> (enqueue '(c . 6))
> (queue->list)
'((c . 6) (b . 8) (a . 12))

> (in-queue? 'b)
#t

> (in-queue? 'x)
#f

> (update-priority 'a 9)
> (queue->list)
'((c . 6) (b . 8) (a . 9))

> (dequeue)
'(c . 6)

> (queue->list)
'((b . 8) (a . 9))

> (peek-queue)
'(b . 8)
```

Regardless of the order the values were added to the queue, they're stored and removed in priority order.

The Implementation

We define our graph as a list of edges. Each edge in the list consists of the end nodes along with the distance between nodes.

```
(define edge-list
  '((S a 12)
    (S b 8)
    (S c 6)
    (a b 1)
    (b c 9)
    (a e 8)
    (e d 5)
    (b d 10)
    (c d 13)))
```

As we progress through the algorithm, we want to to keep track of the current parent of each node so that when the algorithm completes, we'll be able to reproduce the shortest path to each node. A hash table will be used to maintain this information. The key is a node name and the value is the name of the parent node.

```
(define parent (make-hash))
```

We need to take care in our coding to be mindful of the fact that our graph is bi-directional, and an edge defined by (a, b) is equivalent to one defined by (b, a). We'll account for this by supplementing the original edge list with a list consisting of the nodes reversed. We'll also use a hash table (`lengths`) to maintain the lengths of each edge and an additional hash table (`dist`) to record the shortest distance to each node as it's discovered. To pull all this together, we define `init-graph`, which takes an edge list and returns the original list appended with the swapped node list. It will also be used to initialize the priority queue and the various hash tables.

```
(define lengths (make-hash))
(define dist (make-hash))

(define (init-graph start-node edges)
  (let* ([INFINITY 9999]
         [swapped (map (λ (e) (list (second e) (first e) (third e))) edges)]
         [all-edges (append edges swapped)]
         [nodes (list->set (map (λ (e) (first e)) all-edges))])
    (hash-clear! lengths)
    (for ([e all-edges]) (hash-set! lengths (cons (first e) (second e)) (third
      e)))
    (set! queue (make-heap comp))
    (hash-clear! parent)
```

```
    (hash-clear! dist)
    (for ([n nodes])
      (hash-set! parent n null)
      (hash-set! dist n INFINITY)
      (if (equal? n start-node)
          (enqueue (cons start-node 0))
          (enqueue (cons n INFINITY))))
    (hash-set! dist start-node 0)
    all-edges))
```

Here's the code, dijkstra, that actually computes the shortest paths for each node.

```
(define (dijkstra start-node edges)
  (let ([graph (init-graph start-node edges)])
 ➊ (define (neighbors n)
      (filter
       (λ (e) (and (equal? n (first e)) (in-queue? (second e))))
       graph))
 ➋ (let loop ()
      (let* ([u (car (dequeue))])
        (for ([n (neighbors u)])
        ➌ (let* ([v (second n)]
               ➍ [t (+ (hash-ref dist u) (hash-ref lengths (cons u v)))])
    ➎ (when (< t (hash-ref dist v))
            ➏ (hash-set! dist v t)
            ➐ (hash-set! parent v u)
            ➑ (update-priority v t)))))
   ➒ (when (> (heap-count queue) 0) (loop)))))
```

The dijkstra code takes the starting node symbol and edge list as arguments. It then defines graph, which is the original list of edges appended with a list of the edges with the nodes swapped. As mentioned, the init-graph procedure also initializes all the other data structures required for the algorithm to work. A local neighbors function is defined ➊ that takes a node and returns the list of nodes that are adjacent to the node and still in the queue. The main loop starts ➋ and the first step is to pop the first node in the queue and assign its symbol to u. Next, each of its neighbors (v) is processed ➌. For each neighbor, we compute $t = d(u) + l(u, v)$ ➍ (recall that $d(u)$ is the most current distance estimate from the start symbol to u, and $l(u, v)$ is the length of the edge from u to v). We then test whether $t < d(v)$ ➎, and if it passes the test, we do the following:

1. Assign $d(v) = t$ ➏.
2. Assign u as the parent of v ➐.
3. Update the queue with t as the new priority of v ➑.

Finally, we test whether any values remain on the heap, and if so, repeat the process ➒. Once the algorithm completes, parent will contain the parent of each node. All that remains is to chase the chain of parents to the start symbol to determine the shortest path to the node. This is done by the following get-path function:

```
(define (get-path n)
  (define (loop n)
    (if (equal? null n)
        null
        (let ([p (hash-ref parent n)])
          (cons n (loop p)))))
  (reverse (loop n)))
```

The show-paths procedure will print out paths for all the nodes.

```
(define (show-paths)
  (for ([n (hash-keys parent)])
    (printf "  ~a: ~a\n" n (get-path n))))
```

For convenience we define solve, which takes a starting symbol and edge list, calls dijkstra to compute the shortest paths, and prints out the shortest path to each node.

```
(define (solve start-node edges)
  (dijkstra start-node edges)
  (displayln "Shortest path listing:")
  (show-paths))
```

Given our original graph where we defined the edges in edge-list above, and starting symbol S, we generate the solutions as follows:

```
> (solve 'S edge-list)
Shortest path listing:
  S: (S)
  e: (S b a e)
  a: (S b a)
  d: (S b d)
  c: (S c)
  b: (S b)
```

Let's try this slightly more ambitious example in Figure 7-19 (see [4]).

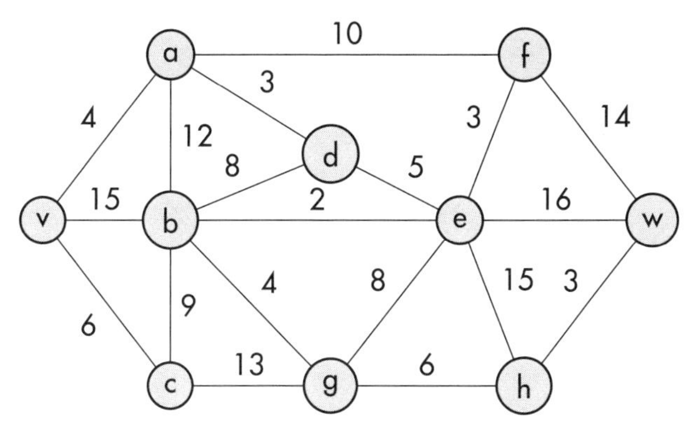

Figure 7-19: Another graph to test Dijkstra's algorithm on

The edge list for this graph is . . .

```
> (define edges '((v a 4) (v b 15) (v c 6) (a b 12) (b c 9) (b d 8) (a f 10) (
    c v 6) (c g 13) (g h 6) (h w 3) (a d 3) (d e 5) (b e 2) (e w 16) (b g 4)
    (g e 8) (e h 16) (e f 3) (f w 14)))
```

So, solving for the shortest path, we have . . .

```
> (solve 'v edges)
Shortest path listing:
  d: (v a d)
  w: (v a d e b g h w)
  f: (v a f)
  c: (v c)
  v: (v)
  a: (v a)
  e: (v a d e)
  g: (v a d e b g)
  h: (v a d e b g h)
  b: (v a d e b)
```

We've highlighted the tree of shortest paths in the resulting graph (see Figure 7-20).

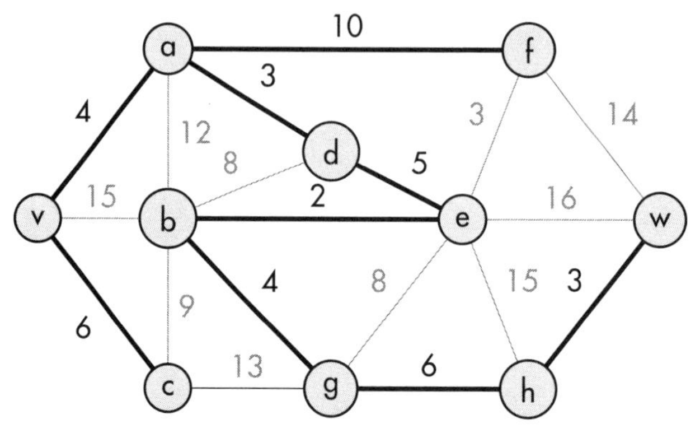

Figure 7-20: Shortest paths found

Now that we've thoroughly examined Dijkstra's shortest-path algorithm, we'll next take a look at the A* algorithm via Sam Loyd's (in)famous 14–15 puzzle.

The 15 Puzzle

The 15 puzzle consists of 15 sequentially numbered sliding tiles in a square frame that are randomly scrambled, with the goal of getting them back into their proper numerical sequence. In the late 1800's Sam Loyd created a bit of buzz over this puzzle by offering a $1,000 prize for anyone who could demonstrate a way to start with a puzzle with all the tiles in order, except with the 14 and 15 tiles reversed (as shown in Figure 7-21, Loyd called this arrangement the "14-15 Puzzle"), and get them back in their proper order (without removing the tiles from the case, of course). As we'll see shortly, this is mathematically impossible, so Loyd knew his money was safe.

Figure 7-21: Sam Loyd's 14–15 puzzle illustration

Why Swapping Just Two Tiles Is Impossible

To get an idea of why Loyd's money was safe (that is, why it's not possible to exchange two and only two tiles), consider the puzzle in its solved state as shown in Figure 7-22.

1	2	3	4
5	6	7	8
9	10	11	12
13	14	15	

Figure 7-22: Solved 15 puzzle

Any sequence of moves that would reverse the 14 and 15 tiles would produce the arrangement in Loyd's puzzle. Simply reproducing this sequence would get the tiles back in order. We'll see that this is impossible. Now consider the arrangement in Figure 7-23.

2	3	1	4
5	6	7	8
9	10	11	12
13	14	15	

Figure 7-23: 15 puzzle with inversions

If we arrange these tiles linearly, we have 2, 3, 1, 4, 5, 6, . . . In particular, the values of tile-2 and tile-3 are larger than tile-1, which follows them. Each such situation, where the value of a tile is larger than one that follows it, is called an *inversion* (two inversions in this case).

Related to the idea of inversion is that of *transposition*. A transposition is simply the exchange of two values in a sequence. A given arrangement can be arrived at by any number of transpositions. For example, one way to get to the sequence 2, 3, 1, 4, 5, 6, . . . would be as follows.

1. Starting arrangement: 1, 2, 3, 4, 5, 6, . . .
2. Transpose 1 and 3: 3, 2, 1, 4, 5, 6, . . .
3. Transpose 2 and 3: 2, 3, 1, 4, 5, 6, . . .

The key idea is that an arrangement consisting of an even number of inversions will always be generated by an even number of transpositions, and an arrangement with an odd number of inversions will always be generated by an odd number of transpositions. For reference purposes, the empty slot will be treated as a tile and designated with the number 16. Any single movement by tile-16 is a transposition. If tile-16 leaves from the lower right corner and arrives at a particular spot using an odd number of transpositions, it will require an odd number of transpositions to get back to the starting location, or a net even number of transpositions. This leaves the puzzle with an even number of inversions.

The arrangement Sam Loyd proposes is impossible to solve since it involves a single odd inversion. It's also true, but trickier to prove, that any puzzle with an even number of inversions is solvable.

Having resolved this historical issue with Sam Loyd's puzzle, we now turn our attention to finding solutions for puzzles that are in fact solvable. In this regard we'll now explore the A* search algorithm (we mostly abbreviate this to simply the "A* algorithm" hereafter).

The A* Search Algorithm

We'll of course assume that the computer is presented with a solvable puzzle (that is, it has an even number of inversions). The computer should provide a solution that's as efficient as possible—that is, a solution that requires the smallest number of moves to get to the goal state. One method that generally provides good results is called the *A* search* algorithm. An advantage of the A* algorithm over a simple breadth-first or depth-first search is that it uses a *heuristic*[1] to reduce the search space. It does this by computing an *estimated* cost of taking any given branch in the search tree. It iteratively improves this estimate until it determines the best solution or it determines no solution can be found. Estimates are stored in a priority queue where the least cost state is at the head of the list.

We'll begin our analysis by looking at a smaller variant of the 15 puzzle, called the 8 puzzle. The 8 puzzle in its solved state is shown in Figure 7-24.

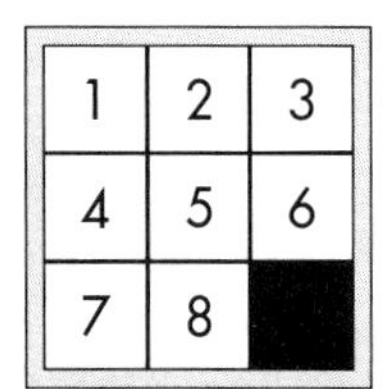

Figure 7-24: Solved 8 puzzle

The search tree of the 8 puzzle can be modeled as shown in Figure 7-25, where each node of the tree is a state of the puzzle, and the child nodes are the possible states that can arise from a valid move.

1. A heuristic is any approach to solving a problem that uses a method that's not guaranteed to offer the best possible solution. Such an approach is often used where the best solution is difficult or impossible to obtain.

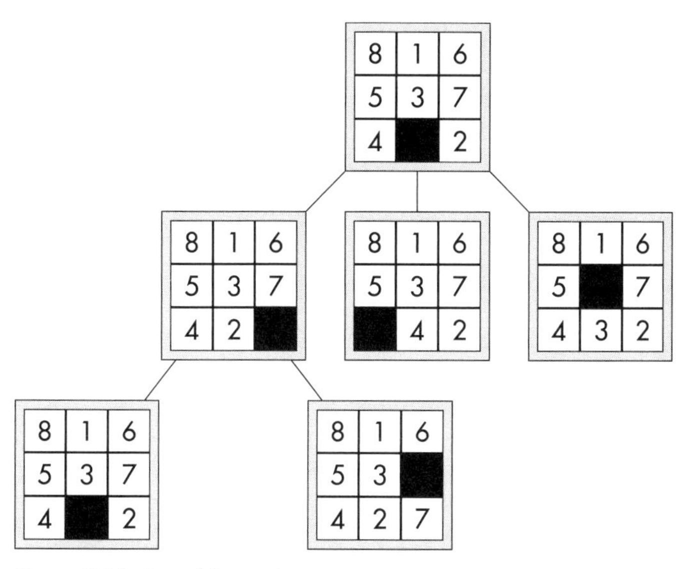

Figure 7-25: Partial 8 puzzle game tree

At each iteration of the A* algorithm, it computes an estimate of the cost (that is, the number of moves) to get to the goal state. Formally, it attempts to minimize the following estimated cost function, where n is the node under consideration, $g(n)$ is the cost of the path from the start node to n, and $h(n)$ is a heuristic that estimates the cost of the cheapest path from n to the goal:

$$f(n) = g(n) + h(n)$$

Designing a good heuristic function is something of an art. To get the best performance from the A* algorithm, an important characteristic of the heuristic is that it satisfies the following condition for every edge in the graph, where $h^*(n)$ is the actual (but unknown) cost to reach the goal state:

$$0 \leq h(n) \leq h^*(n) \quad (7.2)$$

If a heuristic meets this condition, it's said to be *admissible*, and the A* algorithm is guaranteed to find the optimal solution.

One possible heuristic for the 8 puzzle uses something called the *Manhattan distance* (as opposed to the familiar straight-line distance). For example, to get the tile-2 in Figure 7-25 to its home location (the cell it would occupy in the solved state), the tile would have to move up two squares and one square to the left for a total of three moves—this is the Manhattan distance. The heuristic value for a puzzle state would be the sum of the Manhattan distances for each tile. Table 7-1 shows the computation of this value for the root node of Figure 7-25.

Table 7-1: Computing Manhattan Distance

Tile	Rows	Cols.	Total
1	0	1	1
2	2	1	3
3	1	1	2
4	1	0	1
5	0	1	1
6	1	0	1
7	1	2	3
8	2	1	3
		Distance:	15

The Manhattan distance will always be less than or equal to the actual number of moves, so it satisfies the admissibility condition.

A slightly weaker heuristic is the *Hamming distance*, which is just the number of misplaced tiles. The Hamming distance for the puzzle shown in Figure 7-25 is eight: none of the tiles are in their home location.

In Figure 7-26 we've annotated each node with three values. The first value is the depth of the game tree (this value is incremented by one for each level and constitutes the value of $g(n)$ in the cost formula), the second value is the heuristic value, $h(n)$, for the node (the Manhattan distance in this case), and the third value is the sum of the two giving the overall cost score for the node.

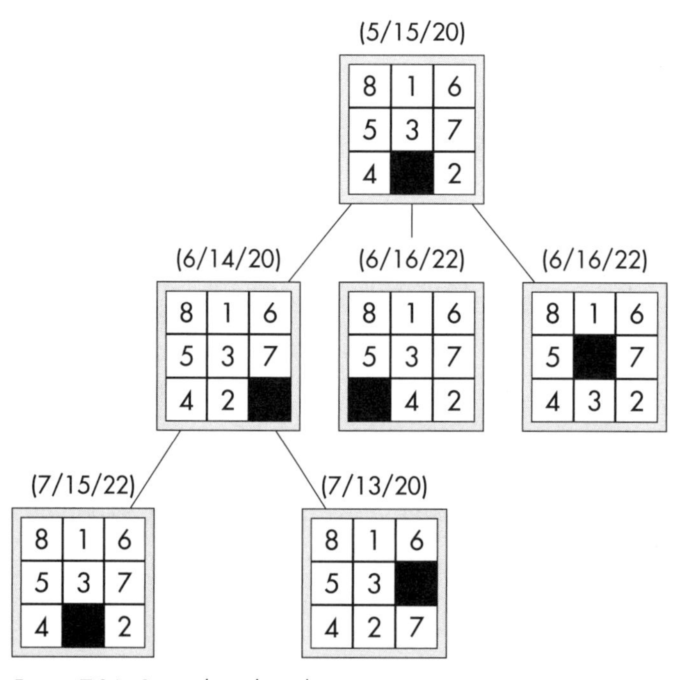

Figure 7-26: 8 puzzle with node costs

The A* algorithm uses a priority queue called *open*. This queue orders the puzzle states that have been examined but whose child nodes have not been expanded, according to the estimated cost to reach the goal. The algorithm also relies on a dictionary, called *closed*, that uses the puzzle state as a key and maintains the most recent cost value for the node. Figure 7-27 reflects the current state of the analysis, where the first node in the open queue is the root node of Figure 7-26.

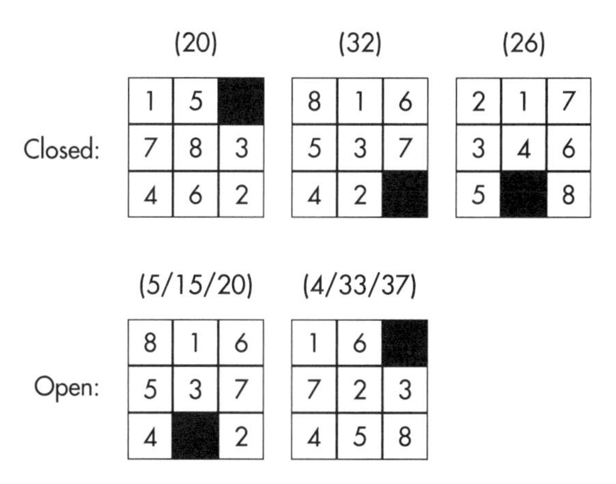

Figure 7-27: Closed and open

The values shown at the top of the closed nodes are the latest estimated costs. The values at the top of the open nodes are the same three cost values described above. With this introduction we walk through an iteration of how the A* algorithm processes the game tree.

The first step is to pop the lowest-priority value off the open queue. This node is then added to the closed dictionary. The next step is to compute the costs for the child nodes, which are shown on the second level of Figure 7-26. If any of these nodes are not on the closed list, they are simply queued to open with no further analysis. Notice that the first child node *is* on the closed list. Since its current estimated cost is less than the cost on the closed list, it's removed from the closed list and the node is added back to the queue with its new value. In the situation where a child node is on the closed list, but its estimated value is *larger* than the value on the closed list, no change is made, and it's not added to the queue. Once this phase is complete, the open and closed structures will appear as shown in Figure 7-28.

Since the first child node in Figure 7-26 has a lower cost than the other nodes in the open queue, it moves to the head of the queue and becomes the next item to be popped off. Notice that its first child is already in the closed list, but its newly computed cost is higher than the cost in the closed list, so it's ignored. The remaining child node would be processed as before.

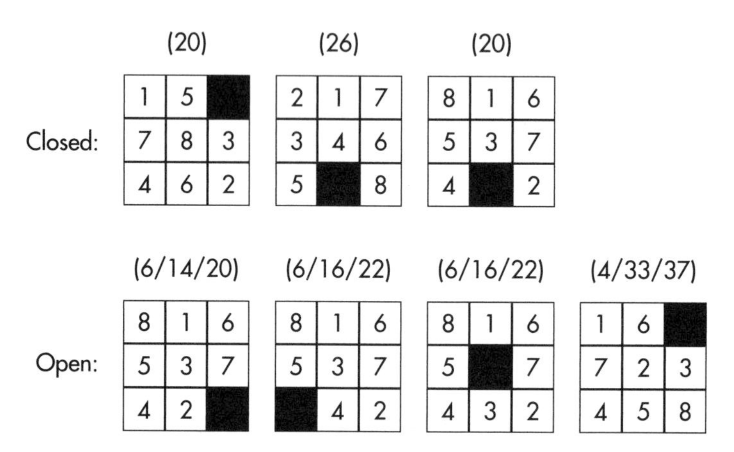

Figure 7-28: Closed and open, updated

The process continues until one of two things happens: either a node is popped off the open queue that's in the solved state, in which case the algorithm completes and prints the answer (details on how this is accomplished are described in the next section), or the open queue becomes empty, which indicates that the puzzle was not solvable.

The 8-Puzzle in Racket

We'll begin by implementing a solution to the smaller 3-by-3 version of the puzzle before we move on the full 4-by-4 version.

As with Dijkstra's algorithm, we'll use Racket's heap object for the open priority queue:

```
#lang racket
(require data/heap)

(define (comp n1 n2)
  (let ([d1 (cdr n1)]
        [d2 (cdr n2)])
    (<= d1 d2)))

(define queue (make-heap comp))

(define (enqueue n) (heap-add! queue n))

(define (dequeue)
  (let ([n (heap-min queue)])
    (heap-remove-min! queue)
    n))
```

A `SIZE` constant will specify the number of rows and columns in the puzzle. In addition we'll define a number of utility functions to work with the puzzle structure. For efficiency, the puzzle state will be stored internally as a Racket vector of size SIZE*SIZE+2. The last two elements of the vector will contain the row and column of the empty cell. The empty cell will have the numerical value specified by (define empty (sqr SIZE)). To this end, we have the following:

```
(define SIZE 3)
(define empty (sqr SIZE))

(define (ref puzzle r c)
  (let ([i (+ c (* r SIZE))])
    (vector-ref puzzle i)))

(define (empty-loc puzzle)
  (values
   (vector-ref puzzle empty)
   (vector-ref puzzle (add1 empty))))
```

The ref function will take a puzzle along with and a row and column number as arguments. It returns the tile number at that location. The empty-loc function will return two values that give the row and column respectively of the empty cell.

The following functions are used to compute the Manhattan distance. The first creates a hash table used to look up the home location of a tile given the tile number. The second function computes the sum of the Manhattan distances for each tile in the puzzle. This will be used in the computation of the cost of a puzzle node.

```
(define tile-homes
  (let ([hash (make-hash)])
    (for ([n (in-range (sqr SIZE))])
      (hash-set! hash (add1 n) (cons (quotient n SIZE) (remainder n SIZE))))
    hash))

(define (manhattan puzzle)
  (let ([dist 0])
    (for* ([r SIZE] [c SIZE])
      (when (not (= empty (ref puzzle r c)))
        (let* ([t (hash-ref tile-homes (ref puzzle r c))]
               [tr (car t)]
               [tc (cdr t)])
          (set! dist (+ dist
                        (abs (- tr r))
                        (abs (- tc c)))))))
    dist))
```

The next functions are used to generate a new puzzle state given a move specifier. A move specifier is a number from zero to three that determines which of the four directions a tile can be moved into the empty space from.

```
(define (move-offset i)
  (case i
    [(0) (values  0 -1)]
    [(1) (values  0  1)]
    [(2) (values -1  0)]
    [(3) (values  1  0)]))

(define (make-move puzzle i)
  (let*-values ([(ro co) (move-offset i)]
                [(re ce) (empty-loc puzzle)]
                [(rt ct) (values (+ re ro) (+ ce co))]
                [(t) (ref puzzle rt ct)])
    (for/vector ([i (in-range (+ 2 (sqr SIZE)))])
      (cond [(< i empty)
             (let-values ([(r c) (quotient/remainder i SIZE)])
               (cond [(and (= r re) (= c ce)) t]
                     [(and (= r rt) (= c ct)) empty]
                     [else (vector-ref puzzle i)]))]
            [(= i empty) rt]
            [else ct]))))
```

The move-offset function takes a move specifier and returns two values specifying the row and column deltas needed to make the move. The make-move function takes a move specifier and returns a new vector representing the puzzle after the move has been made.

The following function will take a puzzle and return a list consisting of all the valid puzzle states that can be reached from a particular puzzle state. The local legal function determines whether a move specifier will result in a valid move for the current puzzle state by checking whether a move in a certain direction will extend beyond the boundaries of the puzzle.

```
(define (next-states puzzle)
  (let-values ([(re ce) (empty-loc puzzle)])
    (define (legal i)
      (let*-values ([(ro co) (move-offset i)]
                    [(rt ct) (values (+ re ro) (+ ce co))])
        (and (>= rt 0) (>= ct 0) (< rt SIZE) (< ct SIZE))))
    (for/list ([i (in-range 4)] #:when (legal i))
      (make-move puzzle i))))
```

It will of course be useful to actually see a visual representation of the puzzle. That functionality is provided by the following routine.

```
(define (print puzzle)
  (for* ([r SIZE] [c SIZE])
    (when (= 0 c) (printf "\n"))
    (let ([t (ref puzzle r c)])
      (if (= t empty)
```

```
        (printf "   ")
        (printf " ~a" t))))
  (printf "\n"))
```

Next we define a helper function to process closed nodes.

```
(define closed (make-hash))

(define (process-closed node-parent node node-depth score)
  (begin
 ❶ (hash-set! closed node (list node-parent score))
    (for ([child (next-states node)])
   ❷ (let* ([depth (add1 node-depth)]
          ❸ [next-score (+ depth (manhattan child))]
          ❹ [next (cons (list child depth node) next-score)])
        (if (hash-has-key? closed child)
         ❺ (let* ([prior-score (second (hash-ref closed child))])
           ❻ (when (< next-score prior-score)
                (hash-remove! closed child)
                (enqueue next)))
         ❼ (enqueue next))))))
```

We begin by placing the node, its parent, and its estimated cost in the `closed` table ❶. Next, we generate a list of the possible child puzzle states and loop through them. For each child node, we generate the new node depth ❷ and estimated score ❸. Then we compile the information that we'd need to push that node to the open queue ❹, which happens automatically ❼ if the node is not in the `closed` table. If the node is in the `closed` table, we extract its prior cost score ❺ and compare with its current score ❻. If the current score is less than the prior score, we remove the node from the `closed` table and place it in the open queue.

We finally get to the algorithm proper.

```
(define (a-star puzzle)
  (let [(solved #f)]
    (hash-clear! closed)
 ❶ (set! queue (make-heap comp))  ; open
 ❷ (enqueue (cons (list puzzle 0 null) (manhattan puzzle)))
❸ (let loop ()
   ❹ (unless solved
        (let* ([node-info (dequeue)])
        ❺ (match node-info
            [(cons (list node node-depth node-parent) score)
          ❻ (if (= 0 (manhattan node))
                (begin
                  (set! solved #t)
                  (print-solution (solution-list node-parent (list node))))
            ❼ (process-closed node-parent node node-depth score)
```

```
               )])
      ❽ (if (> (heap-count queue) 0)
            (loop)
          ❾ (unless solved(printf "No solution found\n"))))))))
```

First we define the closed hash table described above. The open queue is initialized ❶, and the scrambled puzzle provided to the a-star procedure is pushed onto the open queue ❷. The items in the queue consist of a Racket pair. The cdr of the pair is the estimated score, and the car consists of the puzzle state, the depth of the tree, and the parent puzzle state. After this initialization, the main loop starts ❸.

The loop repeats until the solved variable is set to true. The first step of the loop is to pop the highest-priority item (lowest cost score) from the open queue and assign it to the node-info variable. A match form is used to parse the values contained in node-info ❺. The puzzle state (in node) is first tested to see if it's in the solved state ❻, and if so, the function prints out the move sequence and terminates the process. Otherwise, the processing continues where we process the closed node by placing the node, its parent, and estimated cost in the closed table ❼.

Once each iteration completes, the queue is checked ❽ to see if it contains any nodes that needs processing. If it does, then the next iteration resumes ❹; otherwise, no solution exists and the process terminates ❾.

Here are the print functions that show the solution. The solution-list procedure chases the parent nodes in closed to create a list of the puzzle states all the way back to the starting puzzle; print-solution takes the solution list and prints out the puzzle states it contains.

```
(define (solution-list n l)
  (if (equal? n null)
      l
      (let* ([parent (first (hash-ref closed n))])
        (solution-list parent (cons n l)))))

(define (print-solution l)
  (for ([p l]) (print p)))
```

Here's a test run on the puzzle presented in Figure 7-25 (to save space, the output puzzles are displayed horizontally).

```
> (a-star #(8 1 6 5 3 7 4 9 2 2 1))

8 1 6 8 1 6 8 1 6   1 6 1   6 1 3 6 1 3 6
5 3 7 5 3 7   3 7 8 3 7 8 3 7 8   7   8 7
4   2   4 2 5 4 2 5 4 2 5 4 2 5 4 2 5 4 2

1 3 6 1 3 6 1 3 6 1 3 6 1 3 6 1 3 6 1 3 6
5 8 7 5 8 7 5 8 7 5 8   5   8 5 2 8 5 2 8
  4 2 4   2 4 2   4 2 7 4 2 7 4   7 4 7

1 3 6 1 3   1   3 1 2 3 1 2 3 1 2 3 1 2 3
5 2   5 2 6 5 2 6 5   6   5 6 4 5 6 4 5 6
4 7 8 4 7 8 4 7 8 4 7 8 4 7 8   7 8 7   8

1 2 3
4 5 6
7 8
```

Moving Up to the 15 Puzzle

Having laid the groundwork with the 8 puzzle, the 15 puzzle requires this tricky modification: change the value of `SIZE` from 3 to 4. Okay, maybe that's not too tricky, but before you get too excited, it's really not quite that simple either. Observe Figure 7-29. The first two puzzles are from [9], and our test computer sailed through solving these. But the third puzzle was randomly generated and caused the test computer to fall to the floor, giggling to itself that we would attempt to have it solve such a problem—no solution was forthcoming.

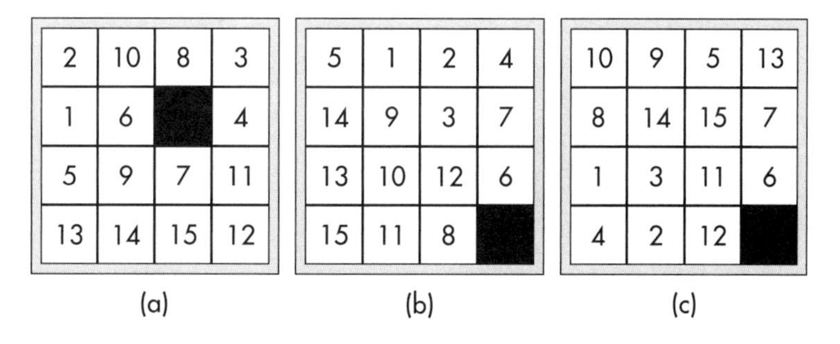

Figure 7-29: Some 15 puzzle examples

The issue is that some puzzles may result in the A* algorithm having to explore too many paths, so the computer runs out of resources (or the user runs out of patience waiting for an answer). Our solution is to settle for a slightly less elegant approach by breaking the problem into three subproblems. We will trade off a fully optimized solution for a solution that we don't have to wait forever to get.

To break the problem into subproblems, we're going to divide the puzzle into three zones as shown below in Figure 7-30. The zones are chosen to take advantage of the fact that the A* algorithm was pretty zippy with the 8 puzzle.

1	2	3	4
5	6	7	8
9	10	11	12
13	14	15	

Figure 7-30: 15 puzzle divided into zones

The medium gray section, which we designate zone 1, represents one subproblem, the white section (zone 2) will be the second subproblem, and the dark gray section (zone 3) represents another subproblem (which is equivalent to the 8 puzzle, which we know can be solved quickly). The idea is to provide different scoring functions to the `a-star` algorithm depending on which zone is being addressed. These functions will still use the Manhattan distance, but with certain restrictions applied. Once zone 1 and zone 2 have been solved, we can just call `manhattan` as before since the edge tiles in zones 1 and 2 will already be in place and the remaining tiles would be equivalent to an 8 puzzle.

Zone 1

Before we dive into solving zone 1, we're going to create a helper function that's quite similar to the code we saw in the `manhattan` function:

```
(define (cost puzzle guard)
  (let ([dist 0])
    (for* ([r SIZE] [c SIZE])
      (let ([t (ref puzzle r c)])
        (when (guard r c t)
          (let* ([th (hash-ref tile-homes t)]
                 [hr (car th)]
                 [hc (cdr th)]
                 [d (+ (abs (- hr r)) (abs (- hc c)))])
            (set! dist (+ dist d))))))
    dist))
```

The main difference is that instead of embedding the test for the empty square directly in the `cost` function, we'll pass in a function for the `guard` parameter that will do the test for us. Given this, we can redefine `manhattan` as follows:

```
(define (manhattan puzzle)
  (cost puzzle (λ (r c t)
                 (not (= t empty)))))
```

We'll attack zone 1 in two phases: first we nudge tiles 1 through 4 into the first two rows of the puzzle; then we get them into the proper order in the first row.

To get tiles 1 through 4 into the first two rows, we define `zone1a`.

```
(define (zone1a puzzle)
  (cost puzzle (λ (r c t)
                 (or (and (<= t 4) (> r 1))))))
```

In this case, we only update the distance for tiles 1 through 4, and we only update the distance if these tiles are not already in the first two rows.

The second phase is just slightly different. This time we always update the distance for tiles 1 through 4 to ensure they land in the proper locations.

```
(define (zone1b puzzle)
  (cost puzzle (λ (r c t)
                 (<= t 4))))
```

It might appear that this could have been done in a single function, but trying to position the tiles all at once would result in a large search space and the consequent increase needed in time and computer resources (like memory). Our first phase, where we just get the tiles into the proximity of their proper location, reduces the search space by half, without requiring an enormous amount of resources. The sorting in phase two will normally only have to deal with the tiles in the top half of the puzzles since the remaining tiles will have a score of zero.

Zone 2

At this point, we've reduced the search space by 25 percent. This modest reduction is sufficient to allow us to get tiles 5, 9, and 13 into zone 2 and in their proper order in a single procedure, which we provide here as `zone2`:

```
(define zone2-tiles (set 5 9 13))

(define (zone2 puzzle)
  (cost puzzle (λ (r c t)
                 (and (>= r 1)
                      (set-member? zone2-tiles t)))))
```

At this point, we no longer need to bother with row 1, as reflected in the code `>= r 1`. Aside from this, the code is nearly identical to the others, except that this time, we use the values defined in `zone2-tiles` for scoring.

Putting It All Together

Once a zone has been solved, we won't want to disturb the tiles that are already in place. To accomplish this, a slight tweak is made to the function (`next-states`) that generates the list of permissible states. Before, we simply checked to ensure we weren't moving beyond the first row or column. Now we define global variables `min-r` and `min-c`, which are set to 0 or 1 depending on which zone we're currently working in.

```
(define min-r 0)
(define min-c 0)

(define (next-states puzzle)
  (let-values ([(re ce) (empty-loc puzzle)])
    (define (legal i)
      (let*-values ([(ro co) (move-offset i)]
                    [(rt ct) (values (+ re ro) (+ ce co))])
        (and (>= rt min-r) (>= ct min-c) (< rt SIZE) (< ct SIZE))))
    (for/list ([i (in-range 4)] #:when (legal i))
      (make-move puzzle i))))
```

The final solver will update the values of `min-r` and `min-c` once zones 1 and 2 have been populated.

We now need to make a few crucial modifications to the code for `a-star` and `process-closed`.

```
(define (process-closed node-parent node node-depth score fscore)
  (begin
    (hash-set! closed node (list node-parent score))
    (for ([child (next-states node)])
      (let* ([depth (add1 node-depth)]
           ❶ [next-score (+ depth (fscore child))]
             [next (cons (list child depth node) next-score)])
        (if (hash-has-key? closed child)
            (let* ([prior-score (second (hash-ref closed child))])
              (when (< next-score prior-score)
                (hash-remove! closed child)
                (enqueue next)))
            (enqueue next))))))
```

The most significant change is that instead of always using `manhattan` for our scoring estimate, we now use the function `fscore` ❶, which is passed to `process-closed` as an additional parameter. This function will be different depending on which zone of the puzzle is being solved.

```
(define (a-star puzzle fscore)
  (let ([solution null]
        [goal null])
    (hash-clear! closed)
    (set! queue (make-heap comp))  ; open
    (enqueue (cons (list puzzle 0 null) (fscore puzzle)))
    (let loop ()
      (when (equal? solution null)
        (let* ([node-info (dequeue)])
          (match node-info
            [(cons (list node node-depth node-parent) score)
             (if (= 0 (fscore node))
                 (begin
               ❶ (set! goal node)
               ❷ (set! solution (solution-list node-parent (list node))))
                 (process-closed node-parent node node-depth score fscore))])
          (if (> (heap-count queue) 0)
              (loop)
              (when (equal? solution null) (printf "No solution found\n"))))))
 ❸ (values goal solution)))
```

Here we also include `fscore` as an additional parameter. Now, instead of immediately printing a solution once the goal state is reached, we return two values at the end ❸: the current goal, given in the first `set!` ❶ and the solution list, given in the second `set!` ❷. The remaining code should align closely with the original version.

Instead of calling `a-star` directly as we did before, we now provide a `solve` function that steps through the process of providing `a-star` with the proper scoring function, which is either one of the zone-specific functions or `manhattan`.

```
(define (solve puzzle)
  (set! min-r 0)
  (set! min-c 0)
  (let*-values ([(goal sol-z1a) (a-star puzzle zone1a)])
    (let*-values  ([(goal sol2) (a-star goal zone1b)]
                   [(sol-z1b) (cdr sol2)])
      (set! min-r 1)
      (let*-values  ([(goal sol3) (a-star goal zone2)]
                     [(sol-z2) (cdr sol3)])
        (set! min-c 1)
        (let*-values  ([(goal sol4) (a-star goal manhattan)]
                       [(sol-man) (cdr sol4)])
          (print-solution (append sol-z1a sol-z1b sol-z2 sol-man)))))))
```

As the code executes, it stores the solution list for each step of the process and finally prints the entire solution in the last line of the code. The reason we take the `cdr` of the solution list at each step (except the first) is because the last item in the solution from the previous step is the goal of the next step; this state would be repeated if we left it in the list.

Finally, we resolve a little cosmetic issue with the `print` routine caused by having two-digit numbers on the tiles. The revised code follows.

```
(define (print puzzle)
  (for* ([r SIZE] [c SIZE])
    (when (= 0 c) (printf "\n"))
    (let ([t (ref puzzle r c)])
      (if (= t empty)
          (printf "   ")
          (printf " ~a" (~a t #:min-width 2 #:align 'right) ))))
  (printf "\n"))
```

The following are some sample inputs with which to test the code. To save space, only the output (compressed) from the first example is shown.

```
> (solve #(2 10 8 3 1 6 16 4 5 9 7 11 13 14 15 12 1 2))

 2 10  8  3    2 10     3    2 10  3        2 10  3  4
 1  6     4    1  6  8  4    1  6  8  4     1  6  8
 5  9  7 11    5  9  7 11    5  9  7 11     5  9  7 11
13 14 15 12   13 14 15 12   13 14 15 12    13 14 15 12

 2 10  3  4    2 10  3  4    2     3  4        2  3  4
 1  6     8    1     6  8    1 10  6  8     1 10  6  8
 5  9  7 11    5  9  7 11    5  9  7 11     5  9  7 11
13 14 15 12   13 14 15 12   13 14 15 12    13 14 15 12

 1  2  3  4    1  2  3  4    1  2  3  4     1  2  3  4
   10  6  8    5 10  6  8    5 10  6  8     5     6  8
 5  9  7 11       9  7 11    9     7 11     9 10  7 11
13 14 15 12   13 14 15 12   13 14 15 12    13 14 15 12

 1  2  3  4    1  2  3  4    1  2  3  4     1  2  3  4
 5  6     8    5  6  7  8    5  6  7  8     5  6  7  8
 9 10  7 11    9 10    11    9 10 11        9 10 11 12
13 14 15 12   13 14 15 12   13 14 15 12    13 14 15

> (solve #(5 1 2 4 14 9 3 7 13 10 12 6 15 11 8 16 3 3))
. . .

> (solve #(10 9 5 13 8 14 15 7 1 3 11 6 4 2 12 16 3 3))
. . .

> (solve #(3 1 2 4 13 6 7 8 5 12 10 11 9 14 15 16 3 3))
```

```
. . .

> (solve #(9 6 12 3 5 13 16 8 14 1 10 7 2 15 11 4 1 2))
. . .

> (solve #(11 1 3 12 5 2 9 8 10 6 14 15 7 13 4 16 3 3))
. . .
```

Be aware that depending on the puzzle and the power of your computer, it may take anywhere from a couple of seconds to a minute or so to generate a solution.

Sudoku

Sudoku[2] is a popular puzzle consisting of a 9-by-9 grid of squares in which some squares are initially populated with digits from 1 through 9, as shown in Figure 7-31a. The objective is to fill in the blank squares such that each row, column, and 3-by-3 block of squares also consists of digits 1 through 9, as shown in Figure 7-31b. A well-formed Sudoku puzzle should only have one possible solution.

1	5					3	2	
6		4		3	1		9	7
			2					
				4	9		7	3
4			8					
								1
5		3			8			
		2						4
		7				2	6	

(a) Start

1	5	8	9	7	4	3	2	6
6	2	4	5	3	1	8	9	7
7	3	9	2	8	6	1	4	5
2	8	5	1	4	9	6	7	3
4	7	1	8	6	3	9	5	2
3	9	6	7	5	2	4	8	1
5	6	3	4	2	8	7	1	9
8	1	2	6	9	7	5	3	4
9	4	7	3	1	5	2	6	8

(b) Solved

Figure 7-31: The Sudoku puzzle

Our aim in this section is to produce a procedure that generates the solution to any given Sudoku puzzle.

The basic strategy is this:

1. Check each empty cell, to determine which numbers are available to be used.
2. Select a cell with the fewest available numbers.
3. One at a time, enter one of the available numbers in the cell.

2. *Sudoku* is the Japanese term for single digit.

4. For each available number, repeat the process until either the puzzle is solved, or there are no numbers available for an empty cell.
5. If there are no available numbers, backtrack to step 3 and try a different number.

The process is another application of depth-first search.

Figure 7-32 gives the coordinates used to reference locations in the puzzle: numbers across the top index columns, numbers on the left edge index rows, and numbers in the interior index blocks of 3-by-3 subgrids.

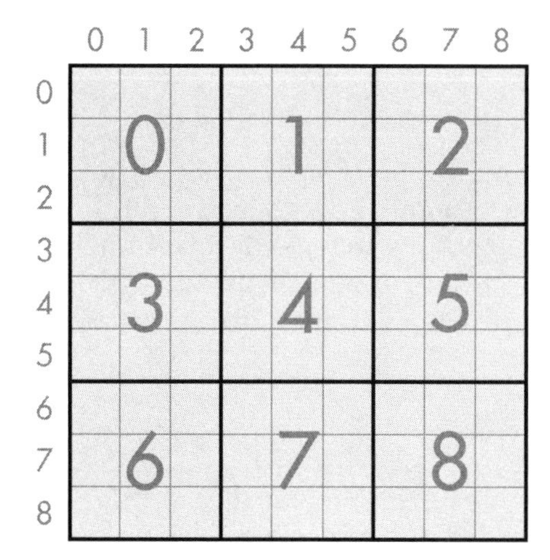

Figure 7-32: Puzzle coordinates

To determine the available numbers for a given cell, we take the set intersection of the unused numbers in each row, column, and block. Using the cell in row 1, column 1 of Figure 7-31a as an example, in row 1 the set of numbers {2, 5, 8} are free, in column 1 all the numbers except for 5 are available, and in block 0, the set of numbers {2, 3, 7, 8, 9} are available. The intersection of these sets gives the set of possible values for this cell: {2, 8}.

Our implementation will use a 9-by-9 array of numbers to represent the puzzle, where the number 0 will designate an empty square. The array will be constructed of a nine-element vector, each element being another nine-element integer vector. For easy access to elements of the array, we define two utility functions to set and retrieve values:

```
(define (array-set! array r c v)
  (vector-set! (vector-ref array r) c v))

(define (array-ref array r c)
  (vector-ref (vector-ref array r) c))
```

Both of these functions require the array and row and column numbers to be provided as the initial arguments.

It will also be useful to derive the corresponding block index from the row and column numbers, as given by the getBlk function here.

```
(define (getBlk r c) ; block from row and column
  (+ (* 3 (quotient r 3)) (quotient c 3)))
```

The puzzle will be input as a single string where each row of nine digits is broken by a new line as shown by the example below.

```
> (define puzzle-str "
150000320
604031097
000200000
000049073
400800000
000000001
503008000
002000004
007000260
")
```

We define the Sudoku puzzle object as a sudoku% Racket object for which we give a partial implementation here. This object will maintain the puzzle state. It will contain functions that allow us to manipulate the state by setting cell values and provides functions to list potential candidates (unused numbers in a row, column, or block) and other helper functions.

```
(define sudoku%
  (class object%

 ❶ (init [puzzle-string ""])

    (define avail-row (make-markers))
    (define avail-col (make-markers))
    (define avail-blk (make-markers))
 ❷ (define count 0)

 ❸ (define grid
      (for/vector ([i 9]) (make-vector 9 0)))

    (super-new)

 ❹ (define/public (item-set! r c n)
      (array-set! grid r c n)
      (array-set! avail-row r n #f)
      (array-set! avail-col c n #f)
      (let ([b (getBlk r c)])
        (array-set! avail-blk b n #f))
      (set! count (+ count 1)))
```

```
    (unless (equal? puzzle-string "")
   ❺ (init-puzzle puzzle-string))

    (define/public (get-grid) grid)

    (define/public (item-ref r c)
      (array-ref grid r c))

  ❻ (define/public (init-grid grid)
      (for* ([r 9] [c 9])
        (let ([n (array-ref grid r c)])
          (when (> n 0)
            (item-set! r c n)))))

  ❼ (define/private (init-puzzle p)
      (let ([g
             (let ([rows (string-split p)])
               (for/vector ([row rows])
                 (for/vector ([c 9])
                   (string->number (substring row c (add1 c))))))])
        (init-grid g)))

; More to come shortly . . .

))
```

The `init` form ❶ captures the input string value that defines the puzzle. We initialize the puzzle ❺ by calling `init-puzzle` ❼, which updates `grid` ❸ with the appropriate numerical values via the call to `init-grid` ❻.

The `count` variable ❷ contains the number of cells that currently have a value. Once `count` reaches 81, the puzzle has been solved.

The `avail-row`, `avail-col`, and `avail-blk` variables are used to keep track of which numbers are currently unused in each row, column, and block respectively. The `make-markers` function, which is called to initialize each of these variables, creates a Boolean array that indicates which numbers are free for any given index (row, column, or block as applicable); `make-markers` is defined as follows:

```
(define (make-markers)
  (for/vector ([i 10])
    (let ([v (make-vector 10 #t)])
      (vector-set! v 0 #f)
      v)))
```

Notice that the number 0 (designating an empty cell) is automatically marked as unavailable.

As numbers get added to the puzzle, `item-set!` is called ❹. This procedure is responsible for updating `grid` and `avail-row`, `avail-col`, and `avail-blk` when given the row, column, and number to be assigned to the puzzle. The functions `get-grid` and `item-ref` return `grid` or a cell in `grid` respectively.

In the following code extracts, all function definitions that are indented should be included in the class definition for `sudoku%` and aren't global defines.

The following `avail` function combines the values from `avail-row`, `avail-col`, and `avail-blk` to produce a vector indicating which numbers are available.

```
(define (avail r c)
  (let* ([b (getBlk r c)]
         [ar (vector-ref avail-row r)]
         [ac (vector-ref avail-col c)]
         [ab (vector-ref avail-blk b)])
    (for/vector ([i 10])
      (and (vector-ref ar i)
           (vector-ref ac i)
           (vector-ref ab i)))))
```

Given this vector, we create a list of the free numbers as follows:

```
(define (free-numbers v)
  (for/list ([n (in-range 1 10)] #:when (vector-ref v n)) n))
```

For efficiency, the following code finds all the cells that only have a single available number and updates the puzzle appropriately.

```
(define (set-singles)
  (let ([found #f])
    (for* ([r 9] [c 9])
      (let* ([free (avail r c)]
             [num-free (vector-count identity free)]
             [n (item-ref r c)])
        (when (and (zero? n)  (= 1 num-free))
          (let ([first-free
                 (let loop ([i 1])
                   (if (vector-ref free i) i
                       (loop (add1 i))))])
            (item-set! r c first-free)
            (set! found #t))
          )))
    found))
```

Performing this process once may result in other cells with only a single available number. The following code runs until no cells remain with only a single available number.

```
(define/public (set-all-singles)
  (when (set-singles) (set-all-singles)))
```

For puzzles that can be solved directly by logic alone (no guessing is required), the above process would be sufficient, but this isn't always the case. To support backtracking, the following two functions are provided.

```
(define (get-free)
  (let ([free-list '()])
    (for* ([r 9] [c 9])
      (let* ([free (avail r c)]
             [num-free (vector-count identity free)]
             [n (item-ref r c)])
        (when (zero? n)
          (set! free-list
                (cons
                 (list r c num-free (free-numbers free))
                 free-list)))))
    free-list))

(define/public (get-min-free)
  (let ([min-free 10]
        [min-info null]
        [free-list (get-free)])
    (let loop ([free free-list])
      (unless (equal? free '())
        (let* ([info (car free)]
               [rem (cdr free)]
               [num-free (third info)])
          (when (< 0 num-free min-free)
            (set! min-free num-free)
            (set! min-info info))
          (loop rem))))
    min-info))
```

The first function (get-free) goes cell by cell and creates a list of all the free values for each cell. Each element of the list contains another list that holds the row, column, number of free values, and a list of the free values. The second function (get-min-free) takes the list returned by get-free and returns the values for the cell with the fewest free numbers.

Here are a few handy utility functions.

```
(define/public (print)
  (for* ([r 9] [c 9])
    (when (zero? c) (printf "\n"))
    (let ([n (item-ref r c)])
      (if (zero? n)
          (printf " .")
          (printf " ~a" n)
          )))
  (printf "\n"))
```

```
(define/public (solved?) (= count 81))

(define/public (clone)
  (let ([p (new sudoku%)])
    (send p init-grid grid)
    p))
```

The print member function provides a simple text printout of the puzzle. The solved? function indicates whether the puzzle is in the solved state by testing whether all 81 cells have been populated. The clone function provides a copy of the puzzle.

This concludes the code that's defined within the body of the sudoku% class definition and brings us to the actual code used to solve the puzzle.

```
(define (solve-sudoku puzzle)
  (let ([solution null]
     ➊ [puzzle (send puzzle clone)])
 ➋ (define (dfs puzzle)
      (if (send puzzle solved?)
          (set! solution puzzle)
          (let ([info (send puzzle get-min-free)])
            (match info
              ['() #f]
           ➌ [(list row col num free-nums)
                (let loop ([nums free-nums])
                  (if (equal? nums '())
                      #f
                   ➍ (let ([n (car nums)]
                           ➎ [t (cdr nums)])
                        (let ([p (send puzzle clone)])
                       ➏ (send p item-set! row col n)
                          (send p set-all-singles)
                       ➐ (unless (dfs p)(loop t))))))]))))
 ➑ (send puzzle set-all-singles)
    (dfs puzzle)
    (if (equal? solution null)
        (error "No solution found.")
        solution
        )))
```

We begin by creating a copy of the puzzle to work with ➊. Next, we define a depth-first search procedure, dfs ➋, which we'll explain shortly. The call to set-all-singles ➑ is occasionally sufficient to solve the puzzle, but the puzzle is handed off to dfs to ensure a complete solution is found. The remaining lines will return a solved puzzle if one exists; otherwise an error is signaled.

The depth-first search code, dfs ➋, immediately tests whether the puzzle is solved and if so, returns the solved puzzle. Otherwise, the cell with the fewest available numbers (if any) is explored ➌, where the match form extracts the cell row, column, number of free numbers, and the list of free numbers. The list of free numbers is iterated through starting on the next line. While the list isn't empty, the first number in the list is extracted into `n` ❹ and the remaining numbers are stored in `t` ❺. Then a copy of the puzzle is created. Next, the puzzle copy is populated with the current available number ❻, and `set-all-singles` is called immediately following this. If this number doesn't produce a solution (via the recursive call to `dfs` ❼), the loop repeats with the original puzzle and the next available number.

To aid testing various puzzles, we define a simple routine to take an input puzzle string, solve the puzzle, and print the solution.

```
(define (solve pstr)
  (let* ([puzzle (new sudoku% [puzzle-string pstr])]
         [solution (solve-sudoku puzzle)])
    (send puzzle print)
    (send solution print)))
```

Now that we've laid this groundwork, here's a trial run with our example puzzle.

```
> (define puzzle "
150000320
604031097
000200000
000049073
400800000
000000001
503008000
002000004
007000260
")
> (solve puzzle)

 1 5 . . . . 3 2 .
 6 . 4 . 3 1 . 9 7
 . . . 2 . . . . .
 . . . . 4 9 . 7 3
 4 . . 8 . . . . .
 . . . . . . . . 1
 5 . 3 . . 8 . . .
 . . 2 . . . . . 4
 . . 7 . . . 2 6 .

 1 5 8 9 7 4 3 2 6
 6 2 4 5 3 1 8 9 7
 7 3 9 2 8 6 1 4 5
 2 8 5 1 4 9 6 7 3
 4 7 1 8 6 3 9 5 2
 3 9 6 7 5 2 4 8 1
```

```
5 6 3 4 2 8 7 1 9
8 1 2 6 9 7 5 3 4
9 4 7 3 1 5 2 6 8
```

While this is certainly an adequate method to generate the output, it doesn't take a lot of additional work to produce a more attractive output.

To accomplish our goal, we'll need the Racket *draw* library.

```
(require racket/draw)
```

Furthermore, we'll borrow our `draw-centered-text` procedure that we used in the 15 puzzle GUI:

```
(define CELL-SIZE 30)

(define (draw-centered-text dc text x y)
  (let-values ([(w h d s) (send dc get-text-extent text)])
    (let ([x (+ x (/ (- CELL-SIZE w) 2))]
          [y (+ y (/ (- CELL-SIZE h d) 2))])
      (send dc draw-text text x y ))))
```

Given these preliminaries, we can now define our `draw-puzzle` function:

```
(define (draw-puzzle p1 p2)
  (let* ([drawing (make-bitmap (* 9 CELL-SIZE) (* 9 CELL-SIZE))]
         [dc (new bitmap-dc% [bitmap drawing])]
         [yellow (new brush% [color (make-object color% 240 210 0)])]
         [gray (new brush% [color "Gainsboro"])])
    (for* ([r 9][c 9])
      (let* ([x (* c CELL-SIZE)]
             [y (* r CELL-SIZE)]
             [n1 (send p1 item-ref r c)]
             [n2 (send p2 item-ref r c)]
             [num (if (zero? n2) "" (number->string n2))]
             [color (if (zero? n1) yellow gray)])
        (send dc set-pen "black" 1 'solid)
        (send dc set-brush color)
        (send dc draw-rectangle x y CELL-SIZE CELL-SIZE)
        (draw-centered-text dc num x y)))
    (for* ([r 3][c 3])
      (let* ([x (* 3 c CELL-SIZE)]
             [y (* 3 r CELL-SIZE)])
        (send dc set-pen "black" 2 'solid)
        (send dc set-brush "black" 'transparent)
        (send dc draw-rectangle x y (* 3 CELL-SIZE) (* 3 CELL-SIZE))))
    drawing))
```

There's really nothing new here. The reason we're passing it two puzzles is because the first puzzle is the original unsolved puzzle. It's simply used to determine which color to use to draw the squares. If a square was blank in the original puzzle, it'll be colored yellow in the output; otherwise it'll be colored gray.

With this in hand, we can redefine solve as follows:

```
(define (solve pstr)
  (let* ([puzzle (new sudoku% [puzzle-string pstr])]
         [solution (solve-sudoku puzzle)])
    (print (draw-puzzle puzzle puzzle))
    (newline)
    (newline)
    (print (draw-puzzle puzzle solution))))
```

Using this new version yields the following:

```
> (solve "
150000320
604031097
000200000
000049073
400800000
000000001
503008000
002000004
007000260
")
```

It gives the starting state in Figure 7-33:

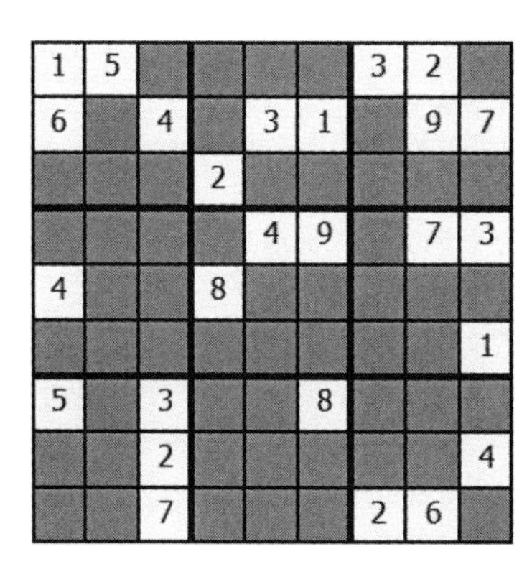

Figure 7-33: Sudoku starting-state drawing

And the solved state in Figure 7-34:

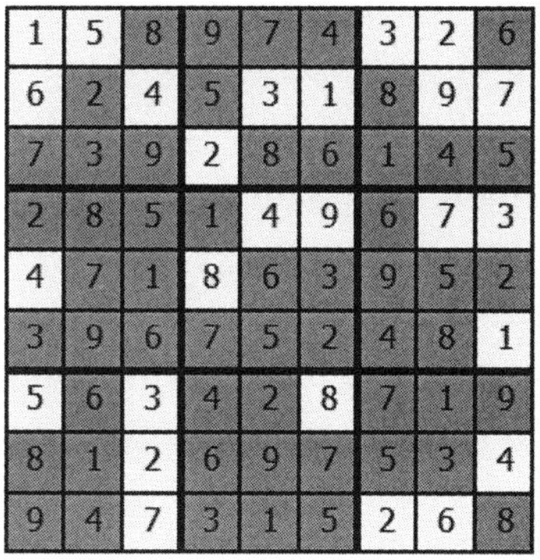

1	5	8	9	7	4	3	2	6
6	2	4	5	3	1	8	9	7
7	3	9	2	8	6	1	4	5
2	8	5	1	4	9	6	7	3
4	7	1	8	6	3	9	5	2
3	9	6	7	5	2	4	8	1
5	6	3	4	2	8	7	1	9
8	1	2	6	9	7	5	3	4
9	4	7	3	1	5	2	6	8

Figure 7-34: Solved Sudoku drawing

Summary

In this chapter, we've explored a number of algorithms that are useful in the general context of problem-solving. In particular we've looked at breath-first search (BFS), depth-first search (DFS), the A* algorithm, and Dijkstra's algorithm (and discovered priority queues along the way) to find the shortest path between graph nodes. We employed DFS in the solution of the *n*-queens problem and the 15 puzzle (which also used the A* algorithm). Finally, we took a look at Sudoku, where sometimes logic alone is sufficient to solve, but failing this, DFS again comes to the rescue. While the algorithms we've explored are far from a comprehensive set, they form a useful toolset that's effective in solving a wide range of problems over many domains.

Thus far we've exercised a number of programming paradigms: imperative, functional, and object-oriented. In the next chapter, we'll look at a new technique: programming in logic or logic programming.

8

LOGIC PROGRAMMING

Logic programming derives its roots from the discipline of formal logic. It's a declarative programming style that focuses on *what* needs to be done, rather than *how* it is to be done. The most well-known programming language in this arena is Prolog (see [5]). The great strength of Prolog, and logic programming, in general is that it provides a platform to express and solve certain types of problems (typically involving some type of search) in a natural and fluid way. The disadvantage is that for other types of problems, logic programming can be very inefficient.

The good news is that Racket allows you to have the best of both worlds. Racket provides a Prolog-style logic programming library called *Racklog*. Racklog closely mirrors Prolog semantics, but as an embedded extension of the Racket syntax. The Racklog library can be accessed via the `(require racklog)` form.

Introduction

Logic programming is all about facts and the relationships between facts. In normal Racket, to define what we consider a coffee drink, we might do it this way:

```
> (define coffee '(moka turkish expresso cappuccino latte))
```

We could then ask whether something is a coffee drink by using the member function.

```
> (member 'latte coffee)
'(latte)

> (member 'milk coffee)
#f
```

The Racklog way of defining our coffee facts is the following. Note that all built-in Racklog object names are prefixed with a percent sign (%) to avoid conflicts with the standard Racket names. Usernames aren't required to use this convention.

```
> (require racklog)

> (define %coffee
    (%rel ()
          [('moka)]
          [('turkish)]
          [('expresso)]
          [('cappuccino)]
          [('latte)]))
```

Such a collection of facts is often referred to as a *database* in Prolog. We can *query* our coffee facts (technically *clauses*) via the `%which` form (asking *which* facts are true). Note that the purpose of the empty parentheses in the `%rel` and `%which` forms will become clear a bit later.

```
> (%which () (%coffee 'latte))
'()

> (%which () (%coffee 'milk))
#f
```

Since `'milk` is not in our `%coffee` facts, the query `(%which () (%coffee 'milk ))` returned false as expected. The expression `(%coffee 'milk)` in the `%which` clause is called the *goal*. Used in this way, `%coffee` is said to be a *predicate*. In essence we're asking, *Is milk a coffee?* In this case, the goal is said to have *failed*. When we asked about `'latte`, our query returned the empty list `'()`. A returned list of any type (even an empty one) is Racklog's way of indicting success. It's possible to query Racklog with explicit goals that always succeed or always fail as follows.

```
> (%which () %true)
'()

> (%which () %fail)
#f
```

Suppose we wanted to know which things are considered a coffee drink. We can ask this way.

```
> (%which (c) (%coffee c))
'((c . moka))
```

When `%which` finds a match, it returns a list of pairs. The `c` identifier is a local logic variable that's used by the `%which` form to indicate which item was matched (that is *bound* or *instantiated*) to the identifier. Note that binding a logic variable is a somewhat different process than binding a Racket identifier. In this case, the identifier `c` isn't assigned a value, but is rather used as a mechanism to associate the logic variable with a value retrieved from the database. While the term *bind* can be used in both senses, we'll usually use the term *instantiate* to distinguish binding a logic variable from binding a Racket identifier. The second subform of `%which` (that is, `(c)`) can be a list of such local logic variables. This list is simply used as a way to declare to Racklog the logic variables that are being used in the remainder of the expression.

What's going on here is a process called *unification*. There are two important factors at work. The first is pattern matching. The second is the aforementioned instantiation. If there are no logic variables in a query, the structure of the query expression must exactly match a corresponding value in the database to succeed. We saw this process fail with the query attempt `(%which () (%coffee 'milk))`, since there was no exact match in the database. If logic variables are part of the query expression, they're allowed to match with corresponding elements in the database. So far, we've only seen a simple example involving a query expression that just consists of a single logic variable where we're querying a database that just contains some atomic values. We'll soon encounter more interesting examples.

We can query our coffee database for more coffee drinks using `(%more)`. Each time `%more` is invoked, additional matches are generated.

```
> (%more)
'((c . turkish))

> (%more)
'((c . expresso))

> (%more)
'((c . cappuccino))

> (%more)
```

```
'((c . latte))

> (%more)
#f
```

Notice that when we run out of coffee facts, (%more) fails (returns #f).

If we just needed to know whether there's any coffee, we could ask this way, where the expression (_) designates an anonymous variable that will match to anything:

```
> (%which () (%coffee (_)))
'()
```

The Basics

What I've shown you so far just looks like a more verbose way to do the same thing Racket already does, but Racklog is built for much more. We'll see that more-complex relationships can be defined, such as parent-child. This type of relationship can be naturally extended to grandparent-child, and so on. Given that these relationships are defined in our database, we can ask questions such as *Who are the parents of Tom?* or *Who are Dick's grandchildren?*

Knowing Your Relatives

Knowing about coffee drinks probably won't keep you awake at night, but knowing who your relatives are just might. That not withstanding, we're going to create a little parent-child database to further expand our knowledge of Racklog.

```
> (define %parent
    (%rel ()
          [('Wilma 'Pebbles)]
          [('Fred 'Pebbles)]
          [('Homer 'Bart)]
          [('Dick 'Harry)]
          [('Sam 'Tim)]
          [('William 'Henry)]
          [('Henry 'John)]
          [('Mary 'Sam)]
          [('Dick 'Harriet)]
          [('Tom 'Dick)]
          [('George 'Sam)]
          [('Tim 'Sue)]))
```

The first item of each relation is the parent, and the second is the child (in reality you can decide which is which; it's just a convention). Suppose that after %parent was defined, it was discovered that 'Lisa and 'Maggie needed to be added as children of 'Homer. This can be remedied by use of one of two %assert! forms.

```
> (%assert! %parent () [('Homer 'Lisa)])
> (%assert-after! %parent () [('Homer 'Maggie)])
```

The first expression adds 'Homer as a parent of 'Lisa after all the other clauses. But be aware that %assert-after! adds a clause *before* all the other clauses (don't ask us). To demonstrate this, let's find all the children of 'Homer.

```
> (%which (c) (%parent 'Homer c))
'((c . Maggie))

> (%more)
'((c . Bart))

> (%more)
'((c . Lisa))
```

It isn't necessary to pre-populate a relation with values. We can create an empty relation and add items to it as shown here.

```
> (define %parent %empty-rel)
> (%assert! %parent () [('Adam 'Bill)])
> (%assert! %parent () [('Noah 'Andy)])
```

We don't need to constrain ourselves to a single generation. We can also ask about grandparents. A grandparent is someone whose child is a parent of someone else. We can define such a relationship in this way:

```
> (define %grand
 ❶ (%rel (g p c)
      ❷ [(g c)
          ❸ (%parent g p) (%parent p c)]))
```

In this case, the second subform ❶ is a list of symbols (g p c) (representing grandparent, parent, and child respectively). As mentioned with %which, this list is simply a way to declare to Racklog the local logic variables that will be used in the rest of the expression. Unlike other relations where each clause only contained a single expression, in this case the clause has three expressions. If you're familiar with Prolog (don't worry if you're not), this would be expressed as something like this:

```
grand(G,C) :- parent(G,P), parent(P,C).
```

An expression of this type is known as a *rule*. In the Racklog version, we have the expression being matched against ❷. In Prolog terminology, this is known as the *head* of the rule (the Racket code (g c) would be equivalent to grand(G,C) in the Prolog version). Following this, we have two subgoals (known as the rule *body*) that must also be matched ❸. In plain English this is interpreted to mean g is the grandparent of c if g is the parent of p and p is the parent of c.

Let's take a look at what happens with the query (%which (k) (%grand 'Tom k)), which is asking which person (k) is a grandchild of 'Tom. With this query, the local variable g ❸ of our %grand definition is instantiated to the value 'Tom. The variables k and c are tied together (even though neither, as yet, has a concrete value); as mentioned above, the process of associating these variables is called *unification*. Racklog then scans through its parent database (assuming our original set of parents) until it finds an entry where 'Tom is a parent. In this case, there's a record indicating that 'Tom is the parent of 'Dick. So the first subgoal succeeds with the result being that p is instantiated to 'Dick. Now the second subgoal is tested ((%parent p c), which through unification becomes (%parent 'Dick c)). Racklog scans its parent database and finds that 'Dick is the parent of 'Harry, at which point the variable c (and by unification k) is instantiated to the value 'Harry. Executing the query in DrRacket, we indeed arrive at the expected results.

```
> (%which (k) (%grand 'Tom k))
'((k . Harry))
```

If we want to see if 'Tom has any other grandchildren, we can use (%more).

```
> (%more)
'((k . Harriet))
```

At the point that the original match was made for 'Harry, parent (p) was instantiated to 'Dick. What's happening behind the scenes with (%more) is that it's actually triggering a failure of the rule. Racklog then *backtracks* over the goal (%parent p c) and uninstantiates the variable c (it does not uninstantiate p, since this was instantiated in the previous goal). It then looks through the database for another match for a parent of 'Dick and finds a second record with 'Harriet as his child (hence, a grandchild to 'Tom).

One nice feature of logic programming is that the same relation allows questions to be asked in different ways. We've asked who is 'Tom's grandchild, but we could also ask who has grandchildren by framing our query this way:

```
> (%which (g) (%grand g (_)))
'((g . William))

> (%more)
'((g . Tom))

> (%more)
#f
```

Or we could ask whether 'Homer is a grandparent.

```
> (%which () (%grand 'Homer (_)))
#f
```

This is one way in which Racklog extends the capability of Racket. We'll see more of this type of flexibility in the next section.

If we simply wanted to list the parents, we could use the goal (%parent p (_)), followed by entering some number of (%more) commands. It can be a bit cumbersome to have to keep entering (%more) to see if a goal can be re-satisfied. One way around this is to use %bag-of. The %bag-of predicate takes three arguments: a Racket expression we want to return (in this case, just the value of the logic variable p), the goal to test (in this case (%parent p (_))), and a variable to instantiate the computed list of results to (also p). Here's an example.

```
> (%which (p) (%bag-of p (%parent p (_)) p))
'((p Wilma Fred Homer Dick William Henry Mary Sam George Dick Tom Tim))
```

In this case, we just used p as the computed value, but we could dress up the output a bit by forming a query this way (from this we see that the value of logic variable p is cons'd with the literal 'parent to produce the final result).

```
> (%which (p) (%bag-of (cons 'parent p) (%parent p (_)) p))
'((p
   (parent . Wilma)
   (parent . Fred)
   (parent . Homer)
   (parent . Dick)
   (parent . William)
   (parent . Henry)
   (parent . Mary)
   (parent . Sam)
   (parent . George)
   (parent . Dick)
   (parent . Tom)
   (parent . Tim)))
```

Here's a simpler way to get a similar output.

```
> (%find-all (p) (%parent p (_)))
'(((p . Wilma))
  ((p . Fred))
  ((p . Homer))
  ((p . Dick))
  ((p . William))
  ((p . Henry))
  ((p . Mary))
  ((p . Sam))
  ((p . George))
```

```
((p . Dick))
((p . Tom))
((p . Tim)))
```

Using `%bag-of` and `%find-all` will list values in the same order that would result from using `(%more)`. Because of this, some entries may be repeated (such as `'Dick` in this example). To only get unique values, we can use `%set-of` instead.

```
> (%which (p) (%set-of p (%parent p (_)) p))
'((p Wilma Fred Homer Dick William Henry Mary Sam George Tom Tim))
```

In this section, we've covered some of the basic ideas behind logic programming. For a more detailed description of backtracking, unification, and such, refer to the classic (and very approachable) work on the subject in *Programming in Prolog* by Clocksin and Mellish [5].

Racklog Predicates

So far we've explored some of the fundamental capabilities that Racklog offers. Logic programming is a unique paradigm that dictates the need for some specialized tools to make it fully useful. In this section, we'll take a look at a few of those tools.

Equality

We've seen that unification plays a key role in the semantics of logic programming. Racklog provides the equality predicate `%=`, which uses unification in a direct way to test structural equality and implement the instantiation process. The following examples should provide some insight into various ways in which this predicate can be applied.

```
> (%which (a b) (%= '(1 potato sack) (cons a b)))
'((a . 1) (b potato sack))

> (%which (x y) (%= (vector x 5) (vector 4 y)))
'((x . 4) (y . 5))

> (%which (x y) (%= (vector x 5) (list 4 y)))
#f

> (%which () (%= (list 4 5) (list 4 5)))
'()
```

What's happening in the first example is subtle. Note that `(1 potato sack)` is in fact equivalent to `(1 . (potato sack))` and `(cons a b)` is equivalent to `(a . b)`. This means that via unification, `a` gets instantiated to 1 and `b` gets instantiated to `(potato sack)`. The result is `'((a . 1) (b potato sack))`. An instantiation is always shown as a pair, but we see the first element, `(a . 1)`, displayed as a pair and the second element, `(b potato sack)`, shown as a list.

Recall that a list *is* in fact a pair, just displayed a bit differently. In the case of (b potato sack), b is the car of the pair and (potato sack) is the cdr of the pair.

The opposite of %= is %/=, which means not unifiable. Recall that unification is essentially a matching process. Leveraging the last example, observe the following:

```
> (%which (a) (%= (list 4 5) (list 5 a)))
#f

> (%which (a) (%/= (list 4 5) (list 5 a)))
'((a . _))
```

While in the first example it was possible to instantiate the logic variable a to 5, the attempt to match the 4 in the first list to the 5 in the second list caused the unification to fail. In the second example, the unification still failed, but since we used the not-equal predicate, a list was returned with the logic variable a unbound.

Similar to the equality predicate is the *identical* predicate %==. Unlike %=, %== does not do any instantiation. It checks to see whether two expressions are *exactly* the same.

```
> (%which (a b) (%== (list 1 2) (list a b)))
#f

> (%which () (%== (list 1 2) (list 1 2)))
'()
```

The opposite of %== is %/==, which means not identical.

Let

It's sometimes desirable to use local variables in a query to produce an intermediate result and not have those variables shown in the output. The %let predicate provides a way to establish these hidden variables.

```
> (define %friends %empty-rel)
> (%assert! %friends () [('jack 'jill)])
> (%assert! %friends () [('fred 'barny)])
> (%which (pals) (%let (a b) (%bag-of (cons a b) (%friends a b) pals)))
'((pals (jack . jill) (fred . barny)))
```

In this example, the %bag-of predicate creates a cons pair from the result of friends and instantiates it to pals. Here a and b are lexically local to the %let, so only the unified results are passed out of the expression to pals.

Is

The %ispredicate acts a bit differently than the other Racklog predicates. It takes two arguments: the first expression is frequently (but not always) an identifier and the second a normal Racket expression. The %is expression instantiates the result of evaluating the second expression to the first. Normally, all identifiers in the second expression will need to be instantiated before the `%is` expression is evaluated. The `%is` expression can be used to assign a value to the first argument or to test equality.

```
> (%which (val) (%is val (+ 1 (* 2 3 4))))
'((val . 25))

> (%which () (%is 25 (+ 1 (* 2 3 4))))
'()

> (%which () (%is 5 (+ 1 (* 2 3 4))))
#f
```

One difference between `%is` and `%=` is that for `%is`, any logic variables in its second argument generally need to be instantiated, as seen in these examples.

```
> (%which (x y) (%= (list x 5) (list 4 y)))
'((x . 4) (y . 5))

> (%which (x y) (%is (list x 5) (list 4 y)))
#f

> (%which (x y) (%is (list x y) (list 4 5)))
'((x . 4) (y . 5))
```

There are, however, situations where `%is` may be advantageous. See the Racket manual for details.[1]

Arithmetic Comparisons

Racklog uses `%=:=` to test for numeric equality and `%=/=` to test for numeric inequality, but the other predicates are what you'd normally expect.

```
> (%which () (%=:= 1 2))
#f

> (%which () (%=:= 1 1))
'()

> (%which () (%< 1 2))
'()

> (%which () (%>= 5 (+ 2 3)))
'()
```

Note that these comparisons only perform tests without instantiating logic variables, so an expression like `(%which (a) (%=:= a 2))` will fail.

1. *https://docs.racket-lang.org/racklog/racket-w-logic.html?q=%25is#%28part._is%29*

Logical Operators

Racklog supports the expected logical predicates %not, %and, and %or, as shown below. The built-in %fail goal always fails, and the %true goal always succeeds.

```
> (%which () (%not %fail))
'()

> (%which () (%not %true))
#f

> (%which () (%and %true %true %true))
'()

> (%which () (%and %true %fail %true))
#f

> (%which () (%or %true %fail %true))
'()
```

There's also an %if-then-else predicate: when given three goals, if the first goal succeeds, it evaluates the second goal; otherwise, it evaluates the third goal. Here's a little test framework.

```
#lang racket
(require racklog)

(define %spud
  (%rel ()
        [('Russet 'plain)]
        [('Yam 'sweet)]
        [('Kennebec 'plain)]
        [('Sweet 'sweet)]
        [('LaRette 'nutty)]))

(define %spud-taste
  (%rel (tater t taste)
     [(tater t)
         (%if-then-else
            (%spud tater taste)
            (%is t taste)
            (%is t 'unknown))]))
```

The following exchange illustrates %if-then-else in action.

```
> (%which (taste) (%spud-taste 'LaRette taste))
'((taste . nutty))

> (%which (taste) (%spud-taste 'Yam taste))
'((taste . sweet))
```

```
> (%which (taste) (%spud-taste 'broccoli taste))
'((taste . unknown))
```

Because 'broccoli isn't in the %spud database, the last goal is evaluated and 'unknown is instantiated to taste (via t).

Append

We've already seen the standard Racket version of append, which is a function that (typically) takes two lists and returns a list that consists of the two lists concatenated together, as shown below.

```
> (append '(1 2 3) '(4 5 6))
'(1 2 3 4 5 6)
```

This is a one-way street. We only get to ask one question: if I have two lists, what does the resulting list look like if I join the lists together? With the Racklog version we are about to explore, we can also ask these questions:

1. If I have a result list, what other lists can I combine to get this list?
2. If I have a starting list and a resulting list, what list can I join to the starting list to get the resulting list?
3. If I have an ending list and a result list, what list can I join to the start of the ending list to get the result list?
4. If I have three lists, is the third list the result of appending the first two?

Before we explain how Racklog's %append works, let's take a look at a few examples. This first query answers the original question (the result of concatenating two lists).

```
> (%which (result) (%append '(1 2 3) '(4 5 6) result))
'((result 1 2 3 4 5 6))
```

This query answers question two.

```
> (%which (l1) (%append l1 '(4 5 6) '(1 2 3 4 5 6)))
'((l1 1 2 3))
```

This query answers question three.

```
> (%which (l2) (%append '(1 2 3) l2 '(1 2 3 4 5 6)))
'((l2 4 5 6))
```

And this query answers question one.

```
> (%which (lists)
        (%let (l1 l2)
             (%bag-of (list l1 l2)
                      (%append l1 l2 '(1 2 3 4 5 6)) lists)))
'((lists
```

```
(() (1 2 3 4 5 6))
((1) (2 3 4 5 6))
((1 2) (3 4 5 6))
((1 2 3) (4 5 6))
((1 2 3 4) (5 6))
((1 2 3 4 5) (6))
((1 2 3 4 5 6) ()))
```

Generating all possibilities that satisfy certain conditions is one of the strengths of logic programming.

If `%append` weren't already defined in Racklog, it would be easy enough to create it from scratch (adapted from [5]):

```
(define %append
  (%rel (h l l1 l2 l3)
   ❶ [('() l l)]
   ❷ [((cons h l1) l2 (cons h l3))
     ❸ (%append l1 l2 l3)]))
```

So what's going on with our predicate `%append`? It consists of two clauses. The first ❶ simply says that if the first list is empty, the result of concatenating that list to any list `l` is just `l`. The second clause ❷ is more complicated: `((cons h l1) l2 (cons h l3))` is the head of a rule. The head of this rule expects three arguments, each of which is either a list or an uninstantiated variable:

1. The first element of this argument (if a list) is instantiated to `h`, and the rest of the list is instantiated to `l1`.
2. The second argument is instantiated to `l2`.
3. If the third argument is a logic variable, `(cons h l3)` is used to construct the return value from the `h` provided in the first argument and `l3` produced in the recursive call to `%append` ❸. If this argument is a list, its head must match the `h` in the first argument, and the rest of the list is matched to `l3` in the last line ❸.

As we've seen, any one or two of the arguments to `%append` may simply be an uninstantiated variable. Racklog uses its unification process to link concrete values to the appropriate values and uses placeholders to temporarily allocate space for the other variables until a proper instantiation can be made. We consider the case where the first and second arguments are instantiated to explicit lists. Once the unification process is complete ❷, the variables `l1` (instantiated to the tail of the first supplied list) and `l2` (instantiated to the second list) are used to make a recursive call to `%append` ❸ with the expectation that `l3` will be populated by the recursive call with the result of concatenating the now shorter list `l1` with `l2`. Since `(cons h l3)` is used to form the final value, the end result is that both the original supplied lists are concatenated together.

Here's a walk-through where we append '(1) and '(2 3) (for the sake of brevity, we'll use the equal sign (=) to indicated logic variable bindings):

1. The first step is to call (%which (a) (%append '(1) '(2 3) a)).
2. Then comes our first port of call ❶. Since '(1) doesn't match '(), we fall through to the next case.
3. At this point in the code we have h=1, l1='() and l2='(2 3) ❷ (we'll get to l3 later; it's used to construct the returned value).
4. Next is the recursive call ❸. With the instantiated values, this resolves to (%which (l3) (%append '() '(2 3) l3)).
5. We're at our first port of call again ❶, but now the empty lists do match. With l='(2 3) instanciated with l3, we return l3='(2 3).
6. Since we've returned from the recursive call, the logic variable will be back to the values given in step 3; of particular interest is h=1. But now we also have the value returned from the recursive call of l3 ='(2 3). Our code ❷ says that our return value (a) from this stage is constructed from (cons h l3). That's '(1 2 3), the desired end result.

Other instantiation scenarios can be analyzed in a similar fashion.

Member

Another Racket function that has a Racklog equivalent is %member. If we needed to create this function ourselves, one way to do it would be the following:

```
(define %member
  (%rel (x y)
        [(x (cons x (_)))]
        [(x (cons (_) y)) (%member x y)]))
```

Is should be clear that this first checks to see if x is at the start of the list (that is, (cons x (_)) assigns x the value at the head of the list, so it must match the value being searched for); if not, it checks to see if it occurs somewhere in the rest of the list.

Examples:

```
> (define stooges '(larry curly moe))
> (%which () (%member 'larry stooges))
'()

> (%which () (%member 'fred stooges))
#f

> (%find-all (stooge) (%member stooge stooges))
'(((stooge . larry)) ((stooge . curly)) ((stooge . moe)))
```

Racklog Utilities

In this section, we'll look at implementing a few additional predicates in Racklog. These are all common list operations whose implementations show off the capabilities of logic programming and Racklog. We'll only make use of the `%permutation` predicate later (which we'll explain in detail). You can take the remainder as black boxes, meaning that we illustrate what they do and how to use them via provided examples, without providing detailed explanations of the code.

Select

Depending on how `select` is used, it can either pick single items from a list, return a list with an item removed, or return a list with an item inserted. Here's the definition.

```
(define %select
  (%rel (x r h t)
        [(x (cons x t) t)]
        [(x (cons h t) (cons h r))
           (%select x t r)]))
```

And here are some examples.

```
> (%which (r) (%select 'x '(u v w x y z) r)) ; remove 'x from list
'((r u v w y z))

> (%which (s) (%select s '(u v w x y z) '(u v x y z))) ; find value in first
    list that is not in the second
'((s . w))

> (%find-all (s) (%select s '(u v w x y z) (_)))
'(((s . u)) ((s . v)) ((s . w)) ((s . x)) ((s . y)) ((s . z)))

> (%find-all (l) (%select 'a l '(u v w x y z)))
'(((l a u v w x y z))
  ((l u a v w x y z))
  ((l u v a w x y z))
  ((l u v w a x y z))
  ((l u v w x a y z))
  ((l u v w x y a z))
  ((l u v w x y z a)))
```

Subtract

The `%subtract` predicate is designed to remove one set of elements in a list from the set of elements in another list. It leverages the functionality of the `%select` predicate to achieve its result. The implementation is straightforward and should be easy to understand.

```
(define %subtract
  (%rel (s r h t u)
        [(s '() s)]
        [(s (cons h t) r)
             (%select h s u)
             (%subtract u t r)]))
```

The first parameter of the predicate is the source list of items, the second parameter is the list of items to be removed, and the last parameter is the list to be returned.

Here are a few examples illustrating the use of `%subtract`.

```
> (%which (r) (%subtract '(1 2 3 4) '(2 1) r))
'((r 3 4))

> (%which (r) (%subtract '(1 2 3 4) '(3) r))
'((r 1 2 4))

> (%which (t) (%subtract '(1 2 3 4) t '(2)))
'((t 1 3 4))

> (%which (s) (%subtract s '(1 2 4) '(3)))
'((s 1 2 4 3))
```

Permutation

It can sometimes be useful to obtain all the permutations of a given list. To provide a bit of background on how the following predicate works, it's helpful to imagine a simple way of generating all the permutations of a given list. Suppose we have a list of the digits from 1 to 4. It's clear that each digit must, at some point, appear as the first digit in the list. So one approach is to start with four lists, each of these consisting of a single digit from 1 to 4. For each of these lists, we create a corresponding list that contains all the remaining digits as shown below.

```
(1) (2 3 4)
(2) (1 3 4)
(3) (1 2 4)
(4) (1 2 3)
```

We've now made our problem a bit smaller. Instead of having to generate all the permutations of a list of four digits, we now only need a way to generate the permutations of a list of three digits. Of course we're smart enough to know that we can recursively continue this process to work on smaller and smaller lists. All that remains is to just join the parts back together. This is essentially what the `%permutation` predicate does.

Before diving into the code, it's helpful to recall that `%append` can be used not just to append two lists together, but to find all the ways a list can be split into two parts. For example, if we call `%which (l1 l2) '(1 2 3 4)`, one of the possible outputs is '((l1) (l2 1 2 3 4)) (the value of l1 is the empty list). With that bit of background under our belt, here's the predicate (the code is adapted from [5]).

```
(define %permutation
  (%rel (l h t u v w)
    ❶ [('() '())]
    ❷ [(l (cons h t))
         ❸ (%append v (cons h u) l)
         ❹ (%append v u w)
         ❺ (%permutation w t)]))
```

This predicate takes two arguments: a list to permutate and an identifier to instantiate the returned list of permutations to. Let's see what happens when we call (%which (a) (%permutation '(1 2 3 4) a)). Because the list isn't empty, we blow by the first match attempt ❶. Next, we have l='(1 2 3 4) ❷. The rest of the code at this point is used to construct the return value, so we'll come back to that a bit later. The next line is where things get a bit interesting ❸. As mentioned above, the first call to %append with a list as its third argument will yield an empty list and the list '(1 2 3 4). With this result, we have v = '(), h=1, and u='(2 3 4). Moving to the next line, we see that v='() and u='(2 3 4) are instantiated but w is not, so (%append v u w) just binds w to '(2 3 4) ❹. Finally, we generate the permutations of '(2 3 4) and instantiate the result to t ❺. We're now in a position to construct the return value(s) ❷. This will generate all the permutations that start with 1.

So what about the remaining permutations? Once we've exhausted all the permutations starting with 1, via backtracking ❸, we eventually have %append yielding the lists '(1) and '(2 3 4). At this point we have v='(1), h=2, and u='(3 4), so now we have w='(1 3 4) ❹. The process continues as before, now constructing the permutations of lists starting with 2.

Let's look at the different ways we can arrange the four card suits.

```
> (%find-all (s) (%permutation '(♠ ♣ ♡ ♢) s))
'(((s ♠ ♣ ♡ ♢))
  ((s ♠ ♣ ♢ ♡))
  ((s ♠ ♡ ♣ ♢))
  ((s ♠ ♡ ♢ ♣))
  ((s ♠ ♢ ♣ ♡))
  ((s ♠ ♢ ♡ ♣))
  ((s ♣ ♠ ♡ ♢))
  ((s ♣ ♠ ♢ ♡))
  ((s ♣ ♡ ♠ ♢))
  ((s ♣ ♡ ♢ ♠))
  ((s ♣ ♢ ♠ ♡))
  ((s ♣ ♢ ♡ ♠))
  ((s ♡ ♠ ♣ ♢))
  ((s ♡ ♠ ♢ ♣))
  ((s ♡ ♣ ♠ ♢))
  ((s ♡ ♣ ♢ ♠))
```

```
((s ♡ ♢ ♠ ♣))
((s ♡ ♢ ♣ ♠))
((s ♢ ♠ ♣ ♡))
((s ♢ ♠ ♡ ♣))
((s ♢ ♣ ♠ ♡))
((s ♢ ♣ ♡ ♠))
((s ♢ ♡ ♠ ♣))
((s ♢ ♡ ♣ ♠)))
```

By making a small adjustment, we can create a version of `%permutation` that generates all permutations of a certain length by taking an additional parameter, the desired length:

```
(define %permute-n
  (%rel (l h t u v w n m)
        [((_) '() 0) !]
        [(l (cons h t) n)
            (%append v (cons h u) l)
            (%append v u w)
            (%is m (sub1 n))
            (%permute-n w t m)]))
```

The exclamation point (`!`) on the third line is called a *cut*. The cut is a goal that always succeeds, but is used to prevent backtracking across the cut. What this means is that if the goal immediately following the cut fails (via backtracking or any other reason), the cut prevents backtracking to any previous goals. In this case, once we reach a count of zero, there's no need to look for additional, longer permutations. This will make the process a bit more efficient (that is, the predicate will still work properly without it, but without testing additional permutations that are not needed).

Due to the pattern matching done by Racklog, there's no need to have two separate predicates. We can combine these into a single predicate as follows:

```
(define %permute
  (%rel (l h t u v w n m)

        ;permute all
        [('() '())]
        [(l (cons h t))
            (%append v (cons h u) l)
            (%append v u w)
            (%permute w t)]

        ;permute n
        [((_) '() 0) !]
        [(l (cons h t) n)
            (%append v (cons h u) l)
            (%append v u w)
```

```
        (%is m (sub1 n))
        (%permute w t m)]))
```

Here are a couple of examples:

```
> (%find-all (p) (%permute '(1 2 3) p))
'(((p 1 2 3)) ((p 1 3 2)) ((p 2 1 3)) ((p 2 3 1)) ((p 3 1 2)) ((p 3 2 1)))

> (%find-all (p) (%permute '(1 2 3) p 2))
'(((p 1 2)) ((p 1 3)) ((p 2 1)) ((p 2 3)) ((p 3 1)) ((p 3 2)))
```

Now that we've laid the groundwork, let's look at a few applications.

Applications

So far, we've introduced the basic mechanics of logic programming. As interesting as these topics are, we'll now take a look at solving some real world (but recreational) problems. Here we'll see how logic programming provides a framework to solve problems using a declarative style that more directly mirrors the problem constraints.

SEND + MORE = MONEY

The following famous recreational math problem was published in the July 1924 issue of *The Strand Magazine* by Henry Dudeney.

```
      S E N D
  +   M O R E
  -----------
    M O N E Y
```

Each letter represents a different digit in the solution. Problems of this type are variously known as alphametics, cryptarithmetic, cryptarithm, or word addition. While this problem can be solved with just a pencil and paper, we're going to leverage Racket (via Racklog) to solve it instead. We're going to use an approach that's generally frowned on: brute force. This means we're going to generate (almost) all the possible ways we can assign numbers to the letters (it's obvious that M is 1, so we won't bother looking for that value).

In the following code, we use the `%permute-n` predicate defined in the previous section.

```
#lang at-exp racket

(require infix racklog)

(define %permute-n
    ; see previous section
    ...)

❶ (define %check
```

```
    (%rel (S E N D O R Y s1 s2)
          [((list S E N D O R Y))
        ❷ (%is s1 @${S*1000 + E*100 + N*10 + D +
               1000 + O*100 + R*10 + E})
        ❸ (%is s2 @${10000 + O*1000 + N*100 + E*10 + Y})
           (%=:= s1 s2)]))

❹ (define %solve
    (%rel (S E N D M O R Y p)
          [(S E N D M O R Y)
           (%is M 1)
         ❺ (%permute-n '(0 2 3 4 5 6 7 8 9) p 7)
           (%check p)
         ❻ (%= p (list S E N D O R Y))]))
```

The predicate to solve the puzzle is `%solve` ❹. First, it assigns 1 to M as previously discussed. The unique letters (aside from M) used in this puzzle are S, E, N, D, O, R, and Y. The next step is to generate all the possible permutations of `'(0, 2, 3, 4, 5, 6, 7, 8, 9)` ❺ (taken 7 numbers at a time). A call to the predicate `%check` is used to test whether the particular permutation will result in a solution to the puzzle (more on `%check` in a bit). If the current permutation generates a solution, the resulting assignments are returned ❻. Note that if `%check` fails, we backtrack ❺ to generate another permutation.

The code for `%check` is also fairly simple. At the first `%is` statement ❷, we just form the arithmetic sum for the current permutation of `s1` = SEND + MORE (remember M is implicitly 1—here expanded to 1000). At the second `%is` statement ❸, we form the sum `s2` = MONEY. Finally we test whether `s1` = `s2`. Due to the fairly lengthy arithmetic expressions ❷ ❸, we're taking advantage of the *infix* library so that the computation is clear.

We generate the solution as follows.

```
> (%which (S E N D M O R Y) (%solve S E N D M O R Y))
'((S . 9) (E . 5) (N . 6) (D . 7) (M . 1) (O . 0) (R . 8) (Y . 2))
```

Even though we're using a highly inefficient brute-force approach, on a fairly healthy computer, the solution should appear in under a minute.

Fox, Goose, Beans

The fox, goose, and a bag of beans puzzle is an example of a class of puzzles called river crossing puzzles. It's quite old and dates back to at least the 9th century. These types of puzzles are a natural fit for logic programming systems. The narrative of the puzzle goes something like this:

> Once upon a time, a farmer went to the market and purchased a fox, a goose, and a bag of beans. On his way home, the farmer comes to the bank of the river where he left his boat. But his boat is rather small, and the farmer can carry only himself and a single one of his purchases—the fox, the goose, or the bag of the beans. If

left alone, the fox would eat the goose, and the goose would eat the beans.

The farmer's task is to get himself and his purchases (still intact) to the far bank of the river. How does he do it?

While this puzzle is not difficult to solve by hand, it affords us an opportunity to exercise the inherent ability of Racklog to perform a *depth-first search (DFS)*. To get some idea of how this type of search works, imagine you're on a small island and need to get to the lighthouse, but you don't know how to get there and you don't have a map. One way to get to your destination is just to begin driving and every time you get to a fork in the road, carefully record which path you take. You keep going until you either arrive at your destination or you get to a dead end or a place you already visited. If you get to a dead end or a place you already visited, you *backtrack* to the previous fork and take a path you haven't taken before. If you've already tried all the paths at a particular fork, you backtrack to the fork before that. Eventually, if you keep working in this fashion, you'll have tried all the possible paths and arrived at your destination or you'll discover that you're actually on the wrong island (oops).

Assume the farmer travels east and west across the river. Using the DFS strategy, we keep track of which items we've had on each bank as the search progresses. We begin then with a record of all items on the east bank. At any point, we may elect to travel back to the opposite bank without an item, or we may select a single item to carry back to the opposite bank (as long as these movements do not violate the constraints of the puzzle). We must also ensure that the resulting movement does not create an arrangement of items that previously existed. For example, suppose we begin by carrying the goose across the river. We now have two stored states: one with all the items (including the farmer) on the east bank, and one with the fox and beans on the east bank with the farmer and goose on the west bank. At this point the farmer may elect to travel alone back to the east bank, since this generates a new state, but if the farmer (stupidly) carries the goose back to the east bank, this results in a state already seen (the start state) and should not be considered. Play continues in this fashion until the solution is found.

The west bank is designated by the number 0, and the east bank designated by the number 1. A four-element vector is used to keep track of the program state. Each element of the vector will indicate the location (that is, bank) of each character, as indicated in Table 8-1.

Table 8-1: Fox, Goose, Beans State Vector

Index	Character
0	Farmer
1	Fox
2	Goose
3	Beans

We begin by defining which states aren't permissible in a predicate called %rejects.

```
#lang racket
(require racklog)

(define %reject
  (%rel ()
        [(#(0 1 1 1))]
        [(#(0 1 1 0))]
        [(#(0 0 1 1))]
        [(#(1 0 0 0))]
        [(#(1 0 0 1))]
        [(#(1 1 0 0))]))
```

The first rejected state indicates that if the farmer is on bank 0, it's not permissible to have the fox, goose, and beans on bank 1. The remaining states can be analyzed in a similar fashion. Observing the pattern of the numbers, %rejects can be written a bit more succinctly:

```
(define %reject
  (%rel (x y)
        [((vector x y y (_))) (%=/= x y)]
        [((vector x x y y)) (%=/= x y)]))
```

If the farmer moves an item from one bank to the other, it's necessary to toggle both the farmer's bank and the item's bank. This is handled by the toggle-item function, which takes a state vector and an element index and returns a new state vector. Notice that's a normal Racket function and not a Racklog predicate. How this fits in will be shown next.

```
(define (toggle-item s a)
  (for/vector ([i (in-range 4)])
    (let ([loc (vector-ref s i)])
   ❶ (if (or (zero? i) (= i a))
          (- 1 loc)
          loc))))
```

The code (zero? i) tests for the farmer's index (0), and (= i a) checks for the item's index ❶. Recall that for/vector forms a new vector from the results of each item computed in the let body.

The %gen-move predicate below generates moves consisting of each of the four possible types of boat passengers (represented by the numbers 0 through 3 respectively): the farmer alone, or the farmer with a fox, goose, or bag of beans.

```
(define %gen-move
  (%rel (n t s0 s1)
     ❶ [('() s0 s1)
            (%is s1 (cons 0 (toggle-item s0 0))) !]
```

```
❷ [((cons n (_)) s0 s1)
       (%is s1 (cons n (toggle-item s0 n)))]
❸ [((cons (_) t) s0 s1)
       (%gen-move t s0 s1)]))
```

The predicate is initially called with the list '(0 1 2 3) (representing all the items that can be moved) and the current state. It returns a pair with the car indicating the item being moved and the cdr giving the resulting state. We have the situation where there are no items left to move ❶, so the next line simply toggles the state of the farmer. Notice the cut (!): there's no need to generate additional moves, since there's nothing left to move. Next, we have a non-empty list, so we take the head of the list and toggle the state of that item ❷. Finally, we tackle the rest of the list with a recursive call to %gen-move ❸.

As the search progresses, it'll be necessary to ensure that the program doesn't get into an infinite loop by rechecking states that have already been tested. To facilitate this, we maintain a list that contains the states that have already been visited, and pass this list and a state to check to a %check-history predicate. If the state is in the history list, the check will fail.

```
(define %check-history
  (%rel (state h t)
        [(state '())]
        [(state (cons h t))
      ❶ (%is #t (equal? state h)) ! %fail]
        [(state (cons (_) t))
            (%check-history state t)]))
```

Here, we've encountered a previous state, so we fail without backtracking by following the cut with %fail ❶.

Next up is the %gen-valid-move predicate. This predicate is passed the current state and move history. It first generates a potential move and checks whether the items left on the bank after the move form a legitimate combination (that is, the state isn't in the reject list). If so, it then checks whether the current state has been seen before. If not, it returns the move as a valid move.

```
(define %gen-valid-move
  (%rel (state hist move s a left-behind)
        [(state hist move)
            (%gen-move '(0 1 2 3)  state (cons a s))
            (%is left-behind (toggle-item state a))
            (%not (%reject left-behind))
            (%check-history s hist)
            (%is move (cons a s))]))
```

With the previous appetizers under our belt, we now move on to the main course:

```
(define %solve
  (%rel (a s state hist move moves m1 m2)
        [(state (_) moves moves)
          ❶ (%is #t (equal? state #(1 1 1 1))) !]
        [(state hist m1 m2)
          ❷ (%gen-valid-move state hist (cons a s))
          ❸ (%is move (cons a s))
          ❹ (%solve s (cons s hist) (cons move m1) m2)]))
```

The overall strategy is quite simple: generate a valid move and check for the solved state. If we reach a dead-end, Racklog's automatic backtrack mechanism will back up and try another move that doesn't lead to a previous state. The `%solve` predicate is called with the initial state, an empty list (representing the state history), and a list with the moves generated so far (also empty). The final parameter is an identifier to be instantiated to the list of moves solving the puzzle. First we check to see whether the puzzle is in the solved state ❶; if so, we return the move list. If this isn't the case, we get the next move candidate and resulting state ❷ (these are assigned to `move` ❸), which is used to recursively call `%solve` ❹. If a failure is generated by the `%solve` predicate ❹, backtracking occurs. Since `%is` cannot be re-satisfied, backtracking continues back ❷ where another possible solution is generated. The `%solve` predicate returns a pair: the first element is an indicator of the passengers in the boat (see discussion of `%gen-move` for meaning of numbers), and the second is the state of the east bank after the move.

To actually solve the puzzle, we call `%solve` as shown here:

```
> (%which (moves) (%solve #(0 0 0 0) '() '() moves))
'((moves
   (2 . #(1 1 1 1))
   (0 . #(0 1 0 1))
   (1 . #(1 1 0 1))
   (2 . #(0 0 0 1))
   (3 . #(1 0 1 1))
   (0 . #(0 0 1 0))
   (2 . #(1 0 1 0))))
```

In addition to being in reverse order, the output listing leaves a bit to be desired in terms of readability. To get a more intuitive output, we define a couple of new helper procedures. First, we create a Racklog predicate version of the Racket `printf` form that we call `%print`. It takes a format string as its first argument and a value to print as its second. Making this work requires a bit of a trick. The `printf` function can't be called as a Racklog goal since it isn't a predicate. It doesn't return a value, so normal instantiation won't work. The trick is to enclose the `printf` form in a `begin` form (which evaluates expressions in order and returns the value of the last one), where we return #t as its final expression. We can then use %is to instantiate this with the constant #t to create a predicate that always succeeds.

```
(define %print
  (%rel (fmt val)
       [(fmt val) (%is #t (begin (printf fmt val) #t))]))
```

The second helper procedure is a regular Racket function that takes a state vector and a bank number. It returns a list indicating which items are currently on the bank.

```
(define (get-items s b)
  (for/list ([i (in-range 4)] #:when (= b (vector-ref s i)))
    (vector-ref #(Farmer Fox Goose Beans) i)))
```

Given a list of solution moves, %print-moves (see below) will provide two lines of output for each move: the first line will indicate the direction of movement and the passenger(s) of the boat; the second line of output will consist of a list where the first item is the occupants of bank 0 and the second item is the occupants of bank 1. We leave it as a little exercise for the reader to figure out how it works.

```
(define %print-moves
  (%rel (s t i pass dir b0 b1 d)
       [('()) %true]
       [((cons (cons i s) t))
        (%is pass (vector-ref
            #(Farmer Farmer-Fox Farmer-Goose Farmer-Beans) i))
        (%is d (vector-ref s 0))
        (%is dir (vector-ref #( <- -> ) d))
        (%print "~a\n" (list dir pass))
        (%is b0 (get-items s (- 1 d)))
        (%is b1 (get-items s (vector-ref s 0)))
        (%if-then-else
           (%=:= 0 d)
           (%print "~a\n\n" (list b1 b0))
           (%print "~a\n\n" (list b0 b1)))
        (%print-moves t)]))
```

Finally, we have this:

```
(define %print-solution
  (%rel (moves rev-moves)
       [()
            (%print "~a\n\n" (list (get-items #(0 0 0 0) 0) '()))
            (%solve #(0 0 0 0) '() '() moves)
            (%is rev-moves (reverse moves))
            (%print-moves rev-moves)]))
```

The procedure `%print-solution` doesn't take any arguments, but it generates the solution of the puzzle, reverses the list of moves, and calls `%print-moves` to print out the solution. Here's the much more readable end result:

```
> (%which () (%print-solution))
((Farmer Fox Goose Beans) ())

(-> Farmer-Goose)
((Fox Beans) (Farmer Goose))

(<- Farmer)
((Farmer Fox Beans) (Goose))

(-> Farmer-Fox)
((Beans) (Farmer Fox Goose))

(<- Farmer-Goose)
((Farmer Goose Beans) (Fox))

(-> Farmer-Beans)
((Goose) (Farmer Fox Beans))

(<- Farmer)
((Farmer Goose) (Fox Beans))

(-> Farmer-Goose)
(() (Farmer Fox Goose Beans))

'()
```

Recall that the final empty list is Racklog's way of indicating success.

How Many Donuts?

The following problem appeared in the October 27, 2007, "AskMarilyn" column of *Parade* magazine:

> Jack, Janet, and Chrissy meet at their corner coffeehouse and buy half a dozen donuts. Each friend always tells the truth or always lies. Jack says that he got one donut, but Janet says that Jack got two, and Chrissy says that Jack got more than three. On the other hand, all three friends agree that Janet got two. Assuming that each friend got at least one and that no donut was cut and divided, how many donuts did each friend get?

Logic programming systems eat this type of problem for breakfast (donuts, breakfast–funny, eh?), and Racklog is no exception. What's nice about this problem is that its solution in Racklog is mainly just a declarative statement of the facts (embellished with a few helper items).

Here are a few basic definitions; the comments should be sufficient to explain their function.

```
#lang racket
(require racklog)

; Each person can have from one to six donuts
(define %can-have
  (%rel (d)
        [(d) (%member d '(1 2 3 4 5 6))]))

; an alias for equality
(define %has (%rel (n) [(n n)]))

; if a person doesn't have d donuts, they have n donuts
(define %not-have
  (%rel (n d)
        [(n d)
           (%can-have n)
           (%=/= n d)]))
```

The intent here is to determine how many donuts an individual can have, if we say they can't have a certain number (provided as the second argument). Since `%can-have` gives all the donuts a person can have, the statement `(%=/= n d)])` will give all the donuts they can have, excluding the number they can't have.

Now we list each person's statement in two versions (one in case they're telling the truth, and the other in case they're lying). Here we are abbreviating "Chrissy" to "Chris."

```
(define %statement
  (%rel (Jack Janet Chris)

        ; Jack's statements
        [('jack Jack Janet)
            (%has Janet 2) (%has Jack 1)]
        [('jack Jack Janet)
            (%not-have Janet 2) (%not-have Jack 1)]

        ; Janet's statements
        [('janet Jack Janet)
            (%has Janet 2) (%has Jack 2)]
        [('janet Jack Janet)
            (%not-have Janet 2) (%not-have Jack 2)]

        ; Chris's statements
        [('chris Jack Janet)
            (%has Janet 2) (%can-have Jack) (%> Jack 3)]
        [('chris Jack Janet)
```

```
(%not-have Janet 2) (%can-have Jack) (%<= Jack 3)]))
```

Our solver just needs to check each person's statements and see if the total donuts add up to six.

```
(define %solve
  (%rel (Jack Janet Chris)
       [(Jack Janet Chris)
           (%statement 'jack Jack Janet)
           (%statement 'janet Jack Janet)
           (%statement 'chris Jack Janet)
           (%can-have Chris)
           (%is 6 (+ Jack Janet Chris))]))
```

And voilà:

```
> (%which (Jack Janet Chris) (%solve Jack Janet Chris))
'((Jack . 3) (Janet . 1) (Chris . 2))
```

Boles and Creots

Boles and Creots is an old pencil and paper code-breaking game. It's also known as Bulls and Cows, or Pigs and Bulls. A commercial variation, called *Mastermind*, involves codes consisting of colored pegs. Gameplay progresses by one player selecting a secret code (typically a sequence of four or five unique digits or letters). The other player then proposes a guess, to which they're provided a hint consisting of the number of boles (correct digits in the correct position) and the number of creots (correct digits in the wrong position). The players continue exchanging guesses and hints until the guessing player gets all digits in the proper order.

Here we have the computer attempt to guess a number provided by a human player.

The strategy is fairly simple: the guessing player (in this case the Racklog program) keeps a record of each guess and the corresponding number of boles and creots. Candidate guesses are generated (by a brute-force generation of all possible permutations of the digits 0 through 9) where each candidate is tested against previous guesses to see if they yield a consistent number of boles and creots. If a candidate guess isn't found to be inconsistent with previous guesses, it becomes the next guess presented to the user. To see what we mean, suppose play has progressed as shown in Table 8-2 below.

Table 8-2: Boles and Creots in Progress

Guess	Boles	Creots
2359	0	2
1297	2	1

For the next turn, the first candidate guess is 1973. This guess, compared to the first guess in the table, has two correct digits (but in the wrong position), which gives 0 boles and 2 creots. So far, so good; but when compared with the second guess, we have 1 bole and 2 creots, so it's rejected. Suppose our next candidate guess is 9247. This gives 0 boles and 2 creots when compared to the first guess, and 2 boles and 1 creot when compared to the second guess, so it's a good candidate guess. The program guesses 9247, gets a hint from the user, and updates the table with the guess, boles, and creots. The process repeats until someone wins.

To simulate a game between a guessing computer and hinting human, our Racklog program uses a read-evaluate-print loop (REPL) that prints a guess, waits for input (a hint) from the user, reads that input, and evaluates it to form its next guess.

Let's take a look at a sample session before we begin digging into the code. I've decided the number to be guessed is 12345. My response to each guess is a two-digit number representing the number of boles and creots respectively.

```
> (%which () (%repl))

Guess: (3 8 2 1 7)
03

Guess: (8 3 1 0 5)
12

Guess: (8 2 3 5 6)
21

Guess: (8 2 0 3 4)
12

Guess: (8 1 4 5 2)
04

Guess: (1 2 3 4 5)
50
'()
```

The code for the overall process is given below. It relies on a number of supporting processes that are explained in more detail later.

```
(require racklog)

(define DIGITS 5)

(define %repl
  (%rel (digits guess val boles creots)
       [()
```

```
❶ (%is digits (randomize-digits))
❷ (%is #t (begin (set! history '()) #t))
  (%repl digits)]
[(digits)
❸ (%permute-n digits guess DIGITS)
❹ (%consistent? guess)
❺ (%print "\nGuess: ~a\n" guess)
❻ (%= (cons boles creots) (get-input))
❼ (%update-history guess boles creots)
❽ (%if-then-else (%=:= boles DIGITS) ! %fail)]))
```

The constant DIGITS specifies the number of digits to be used for a guess. The %repl predicate implements the read-evaluate-print loop. The %repl code generates a randomized list of digits to be used to generate the guesses ❶, and the history list is cleared ❷. The actual loop starts ❸ where the permutations are generated. Each permutation is tested ❹, and backtracking occurs until an acceptable candidate guess is generated. Once that happens, the user is presented the guess ❺. The user is then prompted to provide the number of boles and creots, with the resulting input parsed ❻. The history list is then updated ❼. Finally, the input is tested on to see if all the digits are correct ❽, in which case a cut (!) is used to terminate the process. Otherwise a failure is generated, which triggers backtracking and additional guesses.

To keep track of prior guesses, a history list is defined. Each element of the list is a three-element list consisting of the following: a guess, the number of boles, and the number of creots. The history list is populated by the %update-history predicate.

```
(define history '())

(define %update-history
  (%rel (guess boles creots)
        [(guess boles creots)
         (%is #t
              (begin
                (set! history (cons (list guess boles creots) history))
                #t))]))
```

As seen above, a guess is represented by a list of digits. We define a score function that, given two lists of digits, compares them and returns the corresponding number of boles and creots in a pair.

```
(define (score c h)
  (let loop ([l1 c] [l2 h] [boles 0] [creots 0])
    (if (equal? l1 null)
        (cons boles creots)
        (let ([d1 (car l1)]
              [d2 (car l2)]
              [t1 (cdr l1)]
              [t2 (cdr l2)])
```

```
        (if (= d1 d2)
            (loop t1 t2 (add1 boles) creots)
            (loop t1 t2 boles (+ creots (if (member d1 h) 1 0))))))))
```

To prevent the program from always starting with the same initial guess, we define a number generator function to create a jumbled set of digits to choose from:

```
(define (randomize-digits)
  (let loop([count 10] [l '()])
    (if (= count 0) l
    (let ([d (random 10)])
      (if (member d l)
          (loop count l)
          (loop (sub1 count) (cons d l)))))))
```

To create guess candidates, we need to generate lists of permutations of our randomized digits. For this purpose, we reuse the `%permute-n` predicate we introduced in an earlier section.

```
(define %permute-n
  (%rel (l h t u v w n m)
        [((_) '() 0) !]
        [(l (cons h t) n)
            (%append v (cons h u) l)
            (%append v u w)
            (%is m (sub1 n))
            (%permute-n w t m)]))
```

A predicate called `%consistent?` takes a guess and tests whether it's consistent (as defined above) with the elements of `history`. It's called with a candidate guess.

```
(define %consistent?
  (%rel (g h hb hc gb gc t)
        [((_) '()) %true]
        [(g (cons (list h hb hc) t))
            (%is (cons gb gc) (score g h))
            (%and (%=:= hb gb) (%=:= hc gc))
            (%consistent? g t)]
        [(g) (%consistent? g history)]))
```

Controlling input and output is the job of `get-input` and `%print`, as given below.

```
(define %print
  (%rel (fmt val)
        [(fmt val)
         (%is #t (begin (printf fmt val) #t))]))
```

```
(define (get-input)
  (let ([val (read (current-input-port))])
    (let-values ([(boles creots) (quotient/remainder val 10)])
      (cons boles creots))))
```

In the opening sections of this chapter, we introduced the logic programming paradigm and various tools and utilities that expand its capabilities. In this section, we looked at a number of puzzles and problems in recreational mathematics that can be solved via logic programming in a natural and declarative way. These problems illustrate the powerful search mechanism that's an inherent feature of logic programming.

Summary

In this chapter, we've given an overview of the logic programming paradigm and looked at a number of interesting applications. We've seen that in addition to Racket's functional and imperative programming capabilities, it's also quite adept at logic programming given its Racklog library. Logic programming (specifically Prolog) is known to be Turing complete. What this means, in simple terms, is that anything that can be computed with a typical imperative programming language can be computed with a logic program. Technically it means that it can be used to simulate a Turing machine (more on this in a bit). That being said, there are problem domains where logic programming isn't going to be optimal. Cases involving extensive numerical calculations or where a well-known imperative algorithm is already available speak against using logic programming. Logic programming particularly shines in search problems such as we saw in the application section and at symbolic calculations such as those involved in theorem proving. The good news with Racket is that you can choose whichever approach best suits the problem at hand.

In the next chapter, we'll take a look at a number of abstract computing machines, such as the aforementioned Turing machine.

9

COMPUTING MACHINES

We're all used to having access to powerful computing devices with complex architectures and instruction sets, but fundamental ideas of computer science are based on far simpler devices. The basic idea is to begin with the simplest possible devices and determine what types of computations are possible. We'll explore three such devices below, ranging from ones that can only perform simple operations like recognizing strings to ones that can carry out any algorithm.

Finite-State Automata

In this section, we'll introduce an abstract machine—a computational model, not a physical machine—called a *finite-state automaton (FSA)* or *finite-state machine (FSM)*. In spite of its impressive name, a finite-state automaton is really something very simple. The entire raison d'être of finite-state automata is to execute conditional expressions.

An FSA has a (metaphorical) tape with a series of symbols as inputs. Each symbol is read exactly once, and the machine then moves to the next symbol in the sequence. An FSM is modeled using a finite number of states and transitions, hence the name finite-state machine. The machine starts in a given initial state and transitions to another state based on the input symbol. It can exist in exactly one state at any given time. Some of the states are *accepting states*, and if the machine finishes in an accepting state, the input string is valid. It's possible that, depending on the input and possible transitions from a state, the machine can't continue, in which case the string is invalid. The key is that for *every* input, there's a *condition* for each transition.

Finite-state automata can be used to model many different types of computations, but in this section, we'll focus on their popular use in computer science as *recognizers*: programs that, given an input string, indicate whether the sequence is valid or invalid. As an example, we'll look at a recognizer that will accept strings of the form "HELLO," "HELLLO," "HELLOOO," and so on, and reject all other strings. It's clear from the nature of this problem that not only does an FSA need to be able to perform computations conditionally, but repetition (and hence, iteration) also comes into play.

There are a number of different ways to represent an FSM program; we'll employ two of them: a *state-transitions diagram* and a *state table*. A state-transitions diagram is a directed graph that describes how the FSM transitions from one state to the next. Each state is indicated by a circle. The initial state is identified by an arrow pointing to one of the states, and the accepting states (there can be more than one) are usually indicated by double circles. States are connected by lines or arcs, each of which is labeled with an input symbol. The FSM transitions from one state to the next based on whether the current input symbol matches the label on one of its exiting arcs. A string is said to be accepted if it eventually reaches one of the accepting states; otherwise, it's rejected.

An FSM for recognizing our "HELLO" strings is shown in Figure 9-1.

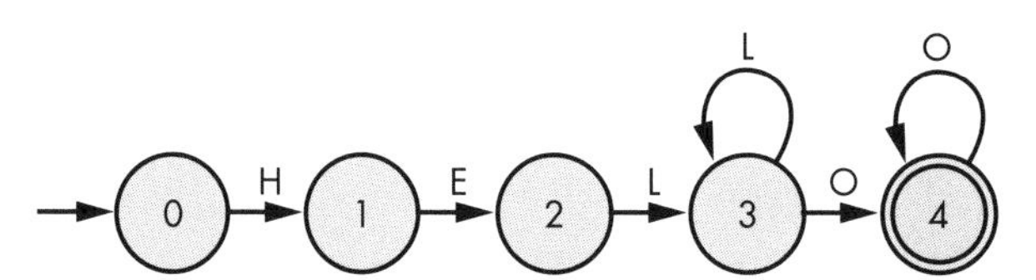

Figure 9-1: An FSM for "HELLO" strings

The FSM begins in state 0, and when it receives an "H" for input, it moves to state 1; otherwise, it halts at state 0 (indicating a rejected string). Once in state 1, it expects an "E" and accepts or rejects the input value as before. It proceeds in this fashion until it reaches the final accepting state (state 4). Observe that state 3 can accept either an "L" or an "O." An FSM, such as this one, built in such a way that every state has only a single transition for each input symbol, is called a *deterministic FSM* (or a *DFA*). It's possible to build finite-state automata where one or more states may transition to more than one state for the same input symbol. Such an FSM is called a *nondeterministic FSM* (or an *NFA*). For any nondeterministic FSM, we can construct a deterministic FSM so they're both capable of recognizing the same set of strings. The advantage of nondeterministic FSMs is that in some cases, they're simpler than their deterministic counterparts. In this text, we'll only be using deterministic finite-state machines.

We can also represent our FSM as a state table (sometimes called an *event table*). See Table 9-1. The first column of the table contains the state number, and the remaining columns represent the input symbols. The table cells contain the next state for a given input symbol. If no state is given, the input symbol is rejected.

Table 9-1: State Table for "HELLO" Strings

S	H	E	L	O
0	1			
1		2		
2			3	
3			3	4
4				4

To implement an FSM to recognize our "HELLO" strings in Racket, we first define a state table as follows:

```
#lang racket

(define state-table
  (vector
   ;        H  E  L  O
   (vector  1 #f #f #f) ; state 0
   (vector #f  2 #f #f) ; state 1
   (vector #f #f  3 #f) ; state 2
   (vector #f #f  3  4) ; state 3
   (vector #f #f #f  4) ; state 4
   ))
```

In this case, we're using `#f` to indicate an invalid transition.

Since we're using vectors for our state table, we need a way to convert characters to indexes. This is done with a hash table, as shown in the following definition.

```
(define chr->ndx
  (make-hash '[(#\H . 0) (#\E . 1) (#\L . 2) (#\O . 3)] ))
```

Given a state number and character, the following `next-state` function gives the next state (or `#f`) from the state table.

```
(define (next-state i chr)
  (if (hash-has-key? chr->ndx chr)
      (vector-ref (vector-ref state-table i)
                  (hash-ref chr->ndx chr) )
      #f))
```

Finally, here's the DFA to recognize our "HELLO" strings.

```
(define (hello-dfa str)
❶ (let ([chrs (string->list str)])
    (let loop ([state 0] [chrs chrs])
   ❷ (if (equal? chrs '()) ; end of string
      ❸ (if (= state 4)
             #t ;if 4, accepting
             #f ;not 4, not accepting
             )
      ❹ (let ([state (next-state state (car chrs))]
               [tail (cdr chrs)])
           (if (equal? state #f) #f ; invalid
            ❺ (loop state tail)))))))
```

First, we convert our string to a list of characters ❶. Then we loop on this list, first checking to see whether all the characters have been read ❷. We then check the state ❸. If we're in an accepting state (in this case, state 4), we return #t indicating an accepted string; otherwise, we return #f. If the entire string hasn't been processed, we get the next state and the rest of the string ❹. We then start the process over again with the remainder of the string ❺. Here are a few sample runs.

```
> (hello-dfa "HELP")
#f

> (hello-dfa "HELLO")
#t

> (hello-dfa "HELLLLLOOOO")
#t

> (hello-dfa "HELLOS")
#f
```

It turns out that finite-state automata (both deterministic and nondeterministic) have certain limitations. For example, it's not possible to build a finite-state machine that recognizes matching parentheses (since at any point, we would need a mechanism to remember how many open parentheses had been encountered). In the next section, we'll introduce the FSM's smarter brother, the Turing machine.

The Turing Machine

The *Turing machine* is an invention of the brilliant British mathematician Alan Turing. In its simplest form, a Turing machine is an abstract computer that consists of the following components.

- An infinite tape of cells that can contain either a zero or a one (arbitrary symbols are also allowed, but we don't use them here)
- A head that can read or write a value in each cell and move left or right (see Figure 9-2)
- A state table
- A state register that contains the current state

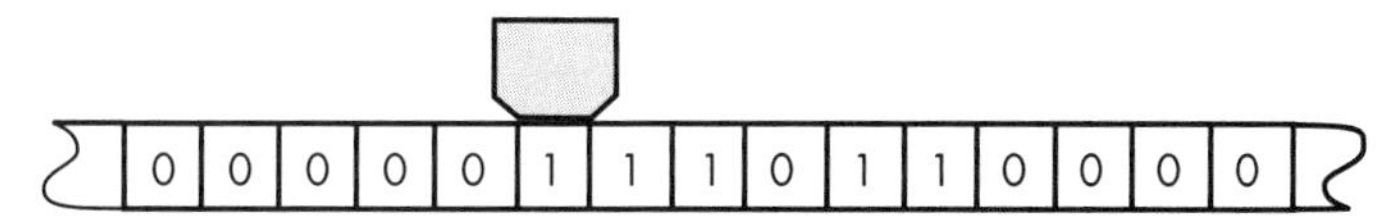

Figure 9-2: Turing machine tape and head

Despite this apparent simplicity, given any computer algorithm, a Turing machine can be constructed that's capable of simulating that algorithm's logic. Conversely, any computing device or programming language that can simulate a Turing machine is said to be *Turing complete*. Thus, a Turing machine is not hampered by the limitations we mentioned for finite-state automata. There's a large body of literature where the Turing machine plays into the analysis of whether certain functions are theoretically computable. We won't engage in these speculations, but rather concentrate on the basic operation of the machine itself.

The machine we construct will have the simple task of adding two numbers. A number, *n*, will be represented as a contiguous string of *n* ones. The numbers to be added will be separated by a single zero, and the head will be positioned at the leftmost one of the first number. The result will be a string of ones, whose length will be the sum of the two numbers. At the end of the computation, the head will be positioned at the leftmost one of the result.

In a nutshell, the program works by changing the leftmost one of the leftmost number to zero, then scanning to the right until it reaches the end of the second number and writing a one just after the last one. The head then moves to the left until one of two things happens:

- It encounters a zero followed by a one, which means there are more ones left to move, so it continues to the left so that it can start over.
- It encounters two consecutive zeros, in which case the addition is done (because there are no other ones left from the first number), and it moves to the right until it is positioned over the leftmost one of the final number.

Figure 9-3 contains some snapshots of the tape at various times during the computation (the triangle shows the head at the start and end of the computation).

▽									
1	1	1	0	1	1	0	0	0	0
0	1	1	0	1	1	0	0	0	0
0	1	1	0	1	1	1	0	0	0
0	0	1	0	1	1	1	0	0	0
0	0	1	0	1	1	1	1	0	0
0	0	0	0	1	1	1	1	0	0
				▽					
0	0	0	0	1	1	1	1	1	0

Figure 9-3: The tale of the tape

You may have already figured out that there are more direct ways to combine the two strings of numbers, but the method described lends itself to being adapted to other computations like multiplication.

Programming a Turing machine consists of constructing a *state table*. Each row in the table represents a particular state. Each state specifies three actions depending on whether the head currently reads a zero or a one. The actions will be to either write a one or a zero to the current cell, whether to move left or right afterward, and what the next state should be. Table 9-2 contains our adder program.

Table 9-2: Turing Machine State Table

	0			**1**		
S	**W**	**M**	**N**	**W**	**M**	**N**
0	x	x	x	0	R	1
1	0	R	2	1	R	1
2	1	L	3	1	R	2
3	0	L	4	1	L	3
4	0	R	6	1	L	5
5	0	R	0	1	L	5
6	0	R	7	x	x	x
7	x	x	x	1	H	x

The topmost row indicates the input symbol. The columns labeled with a W indicate the value that should be written, the columns headed by an M indicate the direction to move (left or right), and the columns headed by an N indicate the next state number. Entries marked with an x are states that will never be reached (this is assuming the inputs and start state are properly set)–in such cases, the entries would be irrelevant. The machine starts in state 0. The final state (or halting state) is state 7, indicted by an H in the move column for a one input.

An alternative (and possibly easier-to-decipher) method of representing a Turing machine's state changes is with a state-transition diagram, as shown in Figure 9-4. In the state diagram, each transition label has three components: the symbol being read, the symbol to write, and the direction to move.

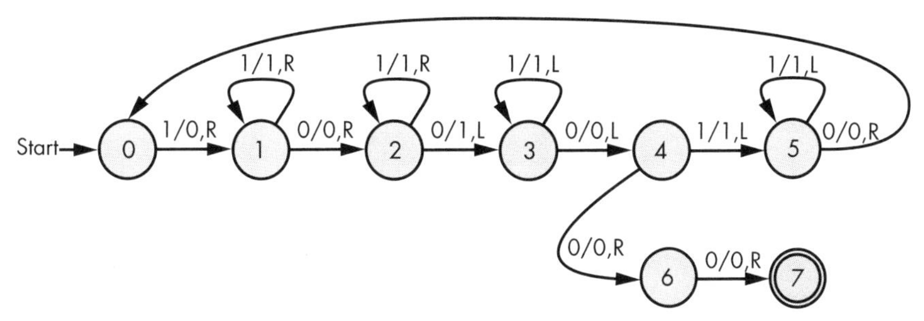

Figure 9-4: Turing machine state-transition diagram

A Racket Turing Machine

As mentioned at the beginning of this section, a programming language that can simulate a Turing machine is said to be Turing complete. We'll demonstrate that Racket is itself Turing complete by constructing just such a simulation (using our addition machine as an example program). We'll of course have to compromise a bit on the infinite tape, so our machine will have a more modest tape, with just 10 cells. The state table will consist of a vector where each cell represents a state. Each state is a two-cell vector where the first cell consists of the actions when a zero is read, and the second cell consists of the actions when a one is read. The actions will be represented by a structure called `act`. The `act` structure will have fields `write`, `move`, and `next` (with obvious meanings). The state will be given in a variable called `state`, and the position of the head will be in `head`. Given these initial considerations, we have the following:

```
#lang racket

(define tape (vector 1 1 1 0 1 1 0 0 0 0))

(define head 0)

(struct act (write move next))

(define state-table
  (vector
   (vector (act 0 #f 0) (act 0 'R 1)) ; state 0
   (vector (act 0 'R 2) (act 1 'R 1)) ; state 1
   (vector (act 1 'L 3) (act 1 'R 2)) ; state 2
   (vector (act 0 'L 4) (act 1 'L 3)) ; state 3
   (vector (act 0 'R 6) (act 1 'L 5)) ; state 4
   (vector (act 0 'R 0) (act 1 'L 5)) ; state 5
```

```
  (vector (act 0 'R 7) (act 1 #f 0)) ; state 6
  (vector (act 0 #f 0) (act 1 'H 0)) ; state 7
  ))

(define state 0)
```

While not strictly required by the definition, we've included an #f value (indicating fail) in the "don't care" states just in case there was some error introduced in the initial setup of the problem (hey, nobody's perfect).

Before getting to the code that specifies the machine's execution, we define a few helper functions to get at various components. The first function returns the next state given the current state and input symbol, and the other two get and set the value at the tape head.

```
(define (state-ref s i) (vector-ref (vector-ref state-table s) i))
(define (head-val) (vector-ref tape head))
(define (tape-set! v) (vector-set! tape head v))
```

The program for running the machine is straightforward. Note that this code is the same for *any* Turing machine you program; only the tape and state-table change.

```
(define (run-machine)
  (let* ([sym (head-val)] ; current input
       ❶ [actions (state-ref state sym)]
         [move (act-move actions)])
    (cond [(equal? #f move)
           (printf "Failure in state ~a, head: ~a\n~a" state head tape)]
        ❷ [(equal? 'H move)
           (printf "Done!\n")]
          [else
         ❸ (let* ([write (act-write actions)]
                ❹ [changed (not (equal? sym write))])
             (tape-set! write)
           ❺ (set! head (if (equal? move 'L) (sub1 head) (add1 head)))
           ❻ (when changed (printf "~a\n" tape))
           ❼ (set! state (act-next actions))
             (run-machine))])))
```

First we capture the actions for the current state and input ❶. Then we capture the next symbol to write ❸ and update the next state ❼. We also test whether the head is about to change a value on the tape ❹, in which case we print out the updated tape ❻. The head position is updated earlier ❺. Once the final state is reached ❷, the program will print Done!. Here's the output.

```
#(1 1 1 0 1 1 0 0 0 0)
#(0 1 1 0 1 1 0 0 0 0)
#(0 1 1 0 1 1 1 0 0 0)
#(0 0 1 0 1 1 1 0 0 0)
#(0 0 1 0 1 1 1 1 0 0)
```

```
#(0 0 0 0 1 1 1 1 0 0)
#(0 0 0 0 1 1 1 1 1 0)
Done!
```

Pushdown Automata

The phrase "pushdown automata" is not a call to go out and knock over unsuspecting robots. No, the term *pushdown automaton* (or *PDA*) refers to a class of abstract computing devices that use a *pushdown stack* (or just *stack* to friends). In terms of computing power, pushdown automata lie squarely between that of finite-state automata and Turing machines.

The advantage of a PDA over a finite-state automaton lies in the stack. The stack forms a basic type of memory. Conceptually, a stack is like a stack of plates where you're only allowed to remove a plate from the top of the stack (called a pop) or add (push) a plate to the top. The rest of the stack can only be accessed via adds or removes from the top. To simulate this in Racket, we define a stack as a string of symbols with two operations:

- **Push**. This operation adds a symbol to the top of the stack (front of the string).
- **Pop**. Pop removes the symbol at the top of the stack and returns it.

A PDA is allowed to read the top symbol on the stack, but it has no access to other stack values. Stack values don't necessarily have to match the values used as input symbols.

Like an FSA, a PDA sequentially reads its input and uses state transitions to determine the next state, but a PDA has the requirement that in addition to being in an accepting state, the stack must also be empty for a string to be accepted (but for practical purposes, in the example below we pre-populate the stack with a unique marker to indicate an empty stack). Be aware that pushdown automata come in deterministic and nondeterministic varieties. Further, nondeterministic pushdown automata are capable of performing a wider range of computations.

While we generally try to keep the presentation informal, we're going to provide a formal description of a PDA since you're likely to run across this type of notation if you elect to do further research on abstract computing machines.[1] If you're not familiar with set notation, you may want to jump to "Set Theory" in Chapter 4 to brush up.

Generally, a PDA is defined as a machine $M = (Q, \Sigma, \Gamma, q_0, Z, F, \delta)$, where the following is true:

- Q is a finite set of states.
- Σ is the set of input symbols.
- Γ is the set of possible stack values.

1. See [10] for example.

- $q_0 \in Q$ is the start state.
- $Z \in \Gamma$ is the initial stack symbol.
- $F \subseteq Q$ is the set of accepting states.
- δ is the set of possible transitions.

The set of permissible transitions are then defined by this somewhat intimidating expression (where Γ^* is used to designate all the possible stack strings, and the symbol ϵ is used to represent the empty string, a string with no symbols in it).

$$\delta \subseteq (Q \times (\Sigma \cup \{\epsilon\}) \times \Gamma) \times (Q \times \Gamma^*)$$

This isn't as bad as it looks. It's basically saying that the set of possible transitions is a subset of all possible combinations of states, input symbols, and stack values (that is, all possibilities pre-transition) with all possible states and stack strings (the possibilities post-transition). The first set of values in parentheses represents inputs to the transition function and consists of the following:

- the currents state: $q \in Q$
- the current input symbol: $i \in (\Sigma \cup \{\epsilon\})$ (remember that we use ϵ to indicate that the remaining string can be empty at some point)
- the value at the top of the stack: $s \in \Gamma$

Given these values, a transition defines the next state (the second Q) and the potential new stack values (Γ^*). For any transition, either the stack will be unaltered, new values will be pushed, or a value will be popped from the top.

Changes to the stack are designated by the notation a/b, where a is the symbol at the top of the stack and b is the resulting string at the top of the stack. For example, if we match some input symbol, the value α is at the top of the stack, and we pop this value without replacing it, we designate this by α/ϵ. If we match the input with α at the top of the stack and need to push β to the top of the stack, this would be designated as $\alpha/\beta\alpha$.

Recognizing Zeros and Ones

Let's set the formalities aside for now and look at a simple example. A popular exercise is to build a PDA that recognizes a string of zeros followed by ones such that the string of ones is exactly the same length as the string of zeros. This isn't possible with a finite-state automaton, since it would need to remember how many zeros it had scanned before it started scanning the ones.

The expression 0^n1^n denotes the string format we're looking for (zero repeated n times followed by one repeated n times), and our input alphabet is $\Sigma = \{0, 1\}$. Any other inputs won't be accepted. To recognize this string, we only need to keep track of the number of zeros read so far, so we'll push a zero to the top of the stack whenever a zero is encountered in the input. When a one is encountered, we pop a zero off the stack; if the number of zeros and ones matches, no zeros will remain on the stack at the end of the input. In order to tell when we have popped the last zero from the stack, we'll pre-populate the stack with a special marker, ω. So our stack symbols are the set $\Gamma = \{0, \omega\}$.

Figure 9-5 is the transition diagram for our PDA.

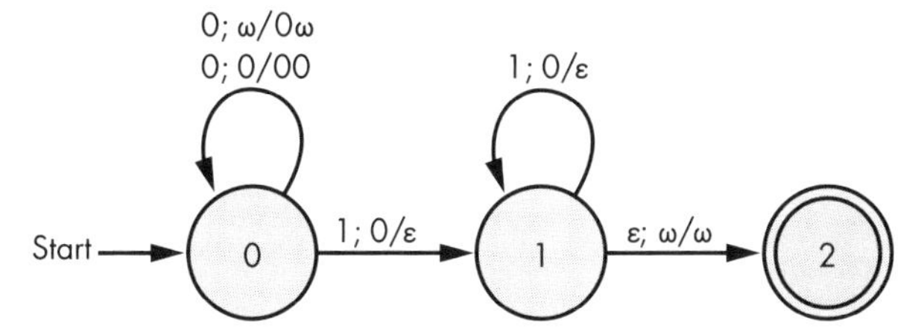

Figure 9-5: Pushdown automaton for 0^n1^n

The label $0; \omega/0\omega$ on the transition looping back to state 0 represents reading a zero on the input with the marker ω at the top of the stack and pushing a zero to the stack. (This is the first transition taken.) Likewise, the label $0; 0/00$ on this loop represents reading a zero on the input with a zero at the top of the stack and pushing a zero to the stack. The label $1; 0/\epsilon$ on the transition from state 0 to state 1 represents reading a one on the input and popping a zero from the stack. The loop on state 1 continues to read ones on the input and pops a zero from the stack for each one read. Once there are no input values and the stack is empty of zeros, the machine moves to state 2, which is an accepting state. Clearly the stack must contain the same number of zeros as the number of ones read in order for the accepting state to be reached.

More Zeros and Ones

Suppose we up the ante a bit and allow any string of zeros and ones, with the only requirement being that there's an equal number of zeros and ones.

Again we assume that the stack is preloaded with ω. This time we allow both zero and one onto the stack. The process is basically this:

- If the top of the stack is ω and there's no more input, the string is accepted.
- If the top of the stack is ω, push the symbol being read.
- If the top of the stack is the same as the symbol being read, push the symbol being read.
- Otherwise, pop the symbol being read.

This process is illustrated by the transition diagram in Figure 9-6.

Neither of these recognizers is possible with finite-state automata. This is due to the fact that in both cases there's a need to remember the number of symbols previously read. The PDA stack (which is not available in a plain FSA) provides this capability.

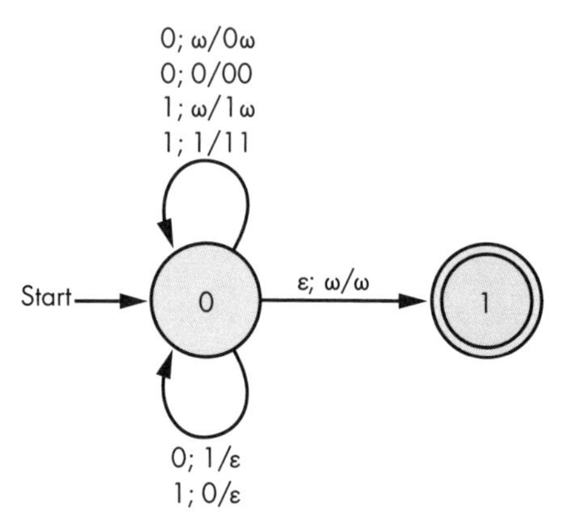

Figure 9-6: PDA to match count of zeros and ones

A Racket PDA

In this section, we'll construct a PDA to recognize the strings described in Figure 9-6. The input will be a list consisting of some sequence of ones and zeros. To process the input, we'll define make-reader, which returns another function.

```
(define (make-reader input)
  (define (read)
    (if (null? input) 'ϵ ; return empty string indicator
        (let ([sym (car input)])
          (set! input (cdr input))
          sym)))
  read)
```

We call make-reader with the list we want to use as input, and it returns a function that will return the next value in the list every time it's called. Here's an example of how it's used.

```
> (define read (make-reader '(1 0 1)))
> (read)
1
> (read)
0
> (read)
1
> (read)
ϵ
```

The stack will also be represented by a list. The following code gives the definitions we need to perform the various stack operations.

```
(define stack '(ω)) ; ω is the bottom of stack marker


(define (pop)
```

```
  (let ([s (car stack)])
    (set! stack (cdr stack))
    s))

(define (push s)
  (set! stack (cons s stack)))

(define (peek) (car stack))
```

Since there's only one state of any significance, we won't bother building a state table. We'll instead take this as an opportunity to exercise another of Racket's hidden treasures, *pattern matching*. This form of pattern matching is Racket's built-in pattern matching as distinct from the pattern matching capability provided by the Racklog library introduced in Chapter 8. Pattern matching uses the `match` form included in the *racket/match* library.[2]

A `match` expression looks a bit like a `cond` expression, but instead of having to use a complex Boolean expression, we simply provide the data structure we want to match against. It's possible to use a number of different structures as patterns to match against, including literal values, but we'll simply use a list for this exercise.

```
(define (run-pda input)
  (let ([read (make-reader input)]) ; initialize the reader
    (set! stack '(ω)) ; initialize stack
    (define (pda)
      (let ([symbol (read)]
            [top (peek)])
        (match (cons symbol top)
          [(cons 'ϵ 'ω ) #t] ; accept input
          [(cons  0 'ω)   (begin (push 0) (pda))]
          [(cons  0  0)  (begin (push 0) (pda))]
          [(cons  1 'ω)   (begin (push 1) (pda))]
          [(cons  1  1)  (begin (push 1) (pda))]
          [(cons  0  1)  (begin (pop)    (pda))]
          [(cons  1  0)  (begin (pop)    (pda))]
          [_ #f]))) ; reject input
    (pda)))
```

Notice how the `match` expression closely mirrors the transitions shown in Figure 9-6. We use `#t` and `#f` to signal whether the input is accepted or rejected. A single underscore (_) serves as a wildcard that matches anything. In this case, matching the wildcard would indicated a rejected string.

Let's take it for a spin.

```
> (run-pda '(1))
#f
```

2. This library is automatically included in the *racket* library.

```
> (run-pda '(1 0))
#t

> (run-pda '(1 0 0 1 1 0))
#t

> (run-pda '(0 1 0 0 1 1 0))
#f

> (run-pda '(1 0 0 1 1 0 0 0 1 1 1))
#f

> (run-pda '(1 0 0 1 1 0 0 0 1 1 1 0))
#t
```

More Automata Fun

Here are a couple of other PDA exercises you may want to try on your own.

- Construct a PDA that matches parentheses (for example, "(())((()))" okay, "(())((())" not okay).
- Build a palindrome recognizer (for example, "madam i'm adam" or "racecar"). This one is tricky and requires constructing a nondeterministic PDA (and also ignoring spaces and punctuation).

A Few Words About Languages

Finite-state automata and pushdown automata serve as recognizers for different classes of languages. A set of symbol strings is called a *regular language* if there's some finite-state machine that accepts the entire set of strings. Examples of regular languages are the set of strings of digits that represent integers, or strings representing floating-point numbers like 1.246e52.

The set of valid arithmetic expressions (for example, $a + x(1 + y)$) is an example of a context-free grammar (CFG). A language consisting of strings accepted by a pushdown automaton is called a context-free language. This means we can construct a pushdown automaton to recognize arithmetic expressions.

Finite-state automata and pushdown automata play a key role in converting modern-day computer language strings into tokens that can then be fed to a PDA to parse the input language. The parser converts the input language into something called an *abstract syntax tree*, which can then be fed to a compiler or interpreter for further processing.

Summary

In this chapter, we explored a number of simple computing machines: finite-state automata, pushdown automata, and the Turing machine. We saw that, while simple, such machines are capable of solving practical problems. In the next chapter, we'll make extensive use of these concepts where their capability to recognize general classes of strings and expressions will be used to develop an interactive calculator.

10

TRAC: THE RACKET ALGEBRAIC CALCULATOR

Racket provides an ecosystem for language-oriented programming. It has extensive built-in capabilities to construct macros, lexers, and parser generators. In this final chapter, we unfortunately won't have time to explore all of these enticing topics. However, we'll explore a number of new topics in computer science and utilize many of the topics introduced in previous chapters (and especially leverage a number of the computing machine concepts introduced in the previous chapter).

In the process, we'll build a command line program called TRAC (The Racket Algebraic Calculator), which will take a string of characters representing an algebraic expression and compute its value. TRAC is, in fact, a stripped-down version of a programming language. If desired, it can be extended in a number of ways to implement a full-fledged programming language.

This program will be able accommodate a dialog such as the following:

```
> let x = 0.8
> let y = 7
> (x + 1.2)*(7.7 / y)
2.2
```

The TRAC Pipeline

To build TRAC, we'll make use of the following pipeline, which processes the input in stages in order to compute the output.

The lexer (or *lexical analyzer*) is responsible for taking the input string and breaking it into a list of tokens that can then be passed to the parser for further processing. Take the following string, for example:

```
"(x + 1.2)*(7.7 / y)"
```

Given the above string, the lexical analyzer will return an output list similar to this:

```
("(" "x" "+" 1.2 ")" "*" "(" 7.7 "/" "Y" ")" )
```

Once we've produced the token list, we can pass it on to the parser. The job of the parser is to determine the structure of the input by building something called the *abstract syntax tree* (or *AST*). An AST is a description of the structure of an expression. Mathematical expressions such as the one just introduced have an inverted tree-like structure. The AST for our example expression is shown in Figure 10-1.

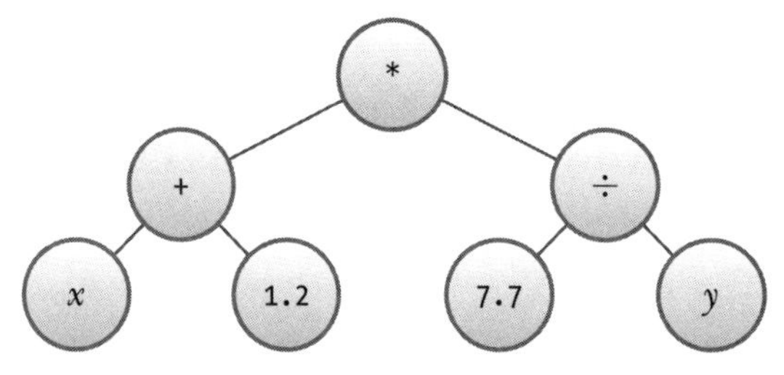

Figure 10-1: AST for (x + 1.2)(7.7 / y)*

We can then pass the AST on to the *interpreter* to evaluate the expression and calculate the result. If we were building a full-blown computer language, the AST would be passed on to a compiler and optimizer, where it would be reduced to machine code for efficient execution. Strictly speaking, if the intent were to only build an interpreter, it wouldn't be necessary to build an AST, since the parser could simply perform any required computations on the fly, but we'll see later that having the AST available will allow us to manipulate it to derive other useful results.

The processing pipeline (lexer, parser, interpreter), in addition to providing a clear separation of duties, allows us to plug in different modules optimized for specific tasks. For example, an interpreter works well for interactive computations but not so much for long running calculations. In such an instance, we'd want to substitute a compiler for the interpreter. This would permit our code to be converted to machine code and run at full speed by being executed directly by the CPU.

We'll discuss and implement each of these components in turn, until we have a working algebraic calculator; then we'll look at a few ways to improve TRAC.

The Lexical Analyzer

In order to split the input into tokens, the lexical analyzer scans the input one character at a time looking for certain patterns. At a high level, a token is just some sequence of characters that can be categorized in a certain way. For example, a string of digits such a 19876 can be categorized as an integer token. Strings of characters that start with a letter and are followed by zero or more letters and digits (such as "AVG1" or "SIN") can be categorized as identifier tokens. Lexical analyzers typically ignore nonessential characters such as spaces and tabs (the language Python is a notable exception).

Each pattern can be represented by a finite-state machine, or FSM (see Chapter 9). One such FSM that we'll use is a recognizer for unsigned integers. In the discussion that follows, certain sets of characters, when grouped together, are referred to as a *character class*. One such class we'll need is the characters consisting of the digits from 0 to 9, which we simply designate as the *digit* class. An unsigned integer is exclusively composed of a string of digits from the digit class, so we can represent its recognizer by the following FSM shown in Figure 10-2, where the digit class is represented by an uppercase italic *D*.

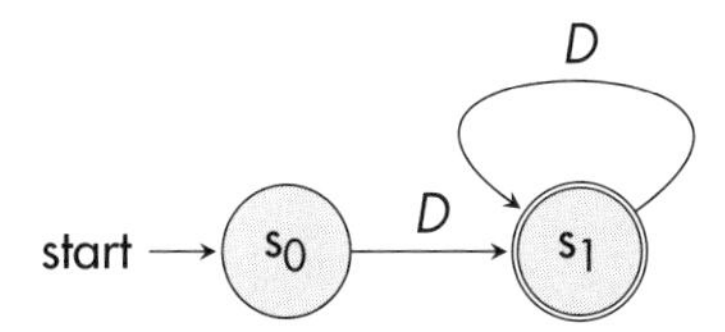

Figure 10-2: FSM to recognize digits

This diagram indicates that an unsigned integer always starts with a digit, and may be followed by any number of trailing digits. An alternative method for representing an unsigned integer is with a *syntax diagram*, such as the one given in Figure 10-3.

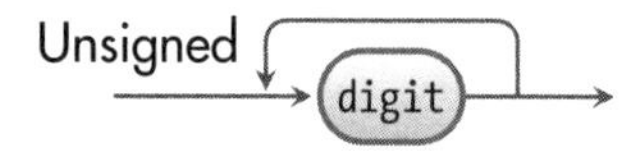

Figure 10-3: Syntax diagram to recognize digits

In this case, the digit class is represented with a typewriter font like this: `digit`. A syntax diagram can sometimes provide a more intuitive representation of the pattern being recognized. The syntax diagram shows that, after accepting a digit, the analyzer can optionally loop back to accept another digit.

To be truly useful, TRAC will need to be able to recognize more than just integers. The following syntax diagram in Figure 10-4 illustrates a recognizer that will accept numbers that consist of unsigned integers, as well as floating-point numbers entered with a decimal point and numbers entered in scientific notation with an embedded `e`.

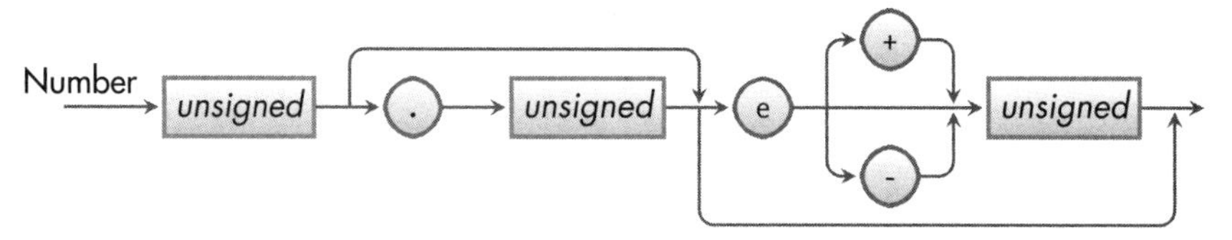

Figure 10-4: Syntax diagram to recognize numbers

Note that syntax diagrams can be nested: the boxes in Figure 10-4 encapsulate the recognizer from Figure 10-3. We leave it as an exercise for the reader to construct the corresponding FSM.

In addition to recognizing numbers, TRAC recognizes identifiers (like `x` in `let x = 4`). TRAC identifiers always start with a letter, followed by any number of letters or digits. We'll designate the letter class with an italic uppercase *L*. As such, the following FSM (see Figure 10-5) will be used to recognize TRAC identifiers:

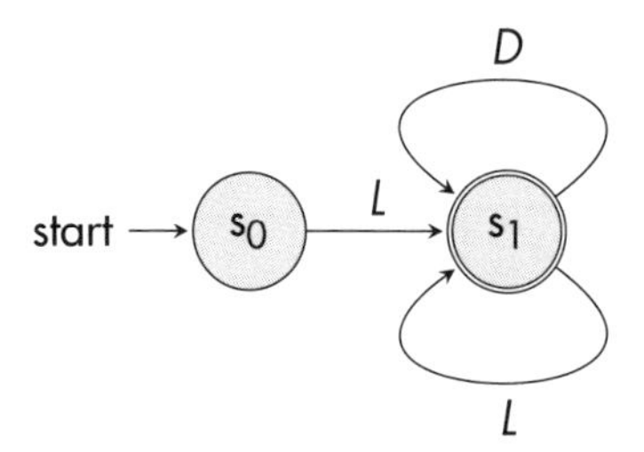

Figure 10-5: FSM to recognize identifiers

Here's the corresponding syntax diagram in Figure 10-6.

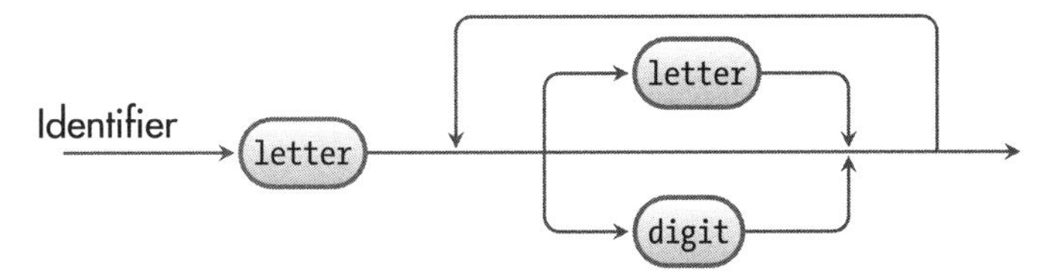

Figure 10-6: Syntax diagram to recognize identifiers

Regular Expressions

So far the discussion has been at a somewhat abstract level. The question now becomes, *How does one actually obtain an FSM that recognizes various character patterns?* The answer is *regular expressions*. A regular expression is essentially a special language used to build finite-state machines (in this case Racket builds the FSM for us, given the regular expression). Our tokens (for example, strings of digits constituting integers) are in fact regular languages. Recall from the last chapter that a regular language is one where there exists an FSM that can accept the entire set of strings. A regular expression is something a bit different. A regular expression (as distinct from a regular language) is really a specification used to build an FSM that recognizes a regular language.

Here's a regular expression that can be used to recognize unsigned integers:

```
[0-9][0-9]*
```

The expression in square brackets is a character class. In this case, it's the class of digits from 0 to 9. This regular expression contains two character classes, both for recognizing digits. The way to interpret this is that the first class will recognize a single digit, but the second, since it's immediately followed by an asterisk, will recognize zero or more additional digits (the asterisk is called the *Kleene star* in honor of Stephen Kleene, who formalized the concept of regular expressions). A more succinct way to do this is with the following regular expression:

```
[0-9]+
```

The trailing plus sign (called the *Kleene plus*) indicates that we want to recognize a string of one or more characters in the class.

The Kleene star and Kleene plus are called *quantifiers*. One additional regular expression quantifier is the question mark, ?. The question mark matches zero or one occurrence of a regular expression. If we wanted to capture numbers that have exactly one or two digits, we could specify it this way:

```
[0-9][0-9]?
```

There are a number of additional ways to specify a regular expression class. The version we've seen for digits specifies a range of values, with the dash (-) separating the start and end characters. It's possible to specify multiple ranges in a class. For example, to specify a class for both the upper- and lowercase characters, one could use [A-Za-z]. A class can also contain any arbitrary set of characters—for example [abCD].

For our purposes, we'll define a class consisting of the arithmetic operators: [-+/*^]. There are a couple of items to note about this particular class. The first is that since the class starts with a dash, the dash isn't used to specify a range, so it's treated as an ordinary character. The second is that the circumflex (^) would be treated differently if it was the first item in the class. For example, the regular expression [^abc] would match all characters *except* a, b, or c.

These are just the basics. Given this overview, let's look at how Racket implements regular expressions and in the process dig deeper into the capability of regular expressions.

Regular Expressions in Racket

Racket builds regular expressions with the regexp function, which takes a string and converts it to a regular expression:

```
> (define unsigned-integer (regexp "[0-9]+"))
```

There's also a special literal regexp value that starts with #rx. For example, a regexp value that recognizes unsigned integers is #rx"[0-9]+" (or #rx "[0-9][0-9]*" if you like typing). This syntax is a shorthand method of constructing regular expressions.

Regexp values are used in conjunction with the functions regexp-match and regexp-match-positions. Suppose we wanted to find the integer embedded in the string "Is the number 1234 an integer?". One way to do it would be with the following:

```
> (regexp-match #rx"[0-9]+" "Is the number 1234 an integer?")
'("1234")
```

The match is returned in a list. The reason for this is that regular expressions can contain subexpressions that will result in additional matches being returned. We'll touch on this a bit later. The regexp-match-positions functions works in a similar fashion to regexp-match. The difference is that regexp-match-positions doesn't return the matched string; instead, it returns the indices that can be used with substring to extract the match. Here's an example.

```
> (regexp-match-positions #rx"[0-9]+" "Is the number 1234 an integer?")
'((14 . 18))

> (substring "Is the number 1234 an integer?" 14 18)
"1234"
```

These functions have a number of useful optional parameters. Instead of searching the entire string, the range can be limited by specifying start and stop positions. Here are some examples.

```
> (regexp-match #rx"[0-9]+" "Is the number 1234 an integer?" 14 18)
'("1234")

> (regexp-match-positions #rx"[0-9]+" "Is the number 1234 an integer?" 14 18)
'((14 . 18))

> (regexp-match #rx"[0-9]+" "Is the number 1234 an integer?" 16)
'("34")
```

Notice in the second example that `regexp-match-positions` always returns the position of the match from the start of the string and not from the specified starting position. The ending position is optional, and if not specified, the search continues until the end of the string is reached, as seen in the third example.

Probably the most basic regular expressions are just literal letters and digits. For example, to determine whether a string contains the string `"gizmo"`, one could form this query:

```
> (regexp-match #rx"gizmo" "Is gizmo here?")
'("gizmo")

> (regexp-match #rx"gizmo" "Is gixmo here?")
#f
```

Of course this type of functionality could be obtained from `string-contains?`, but regular expressions are much more powerful. Used in conjunction with the Kleene star and plus operators, we can form much more sophisticated queries.

```
> (regexp-match #rx"cats.*dogs" "It's raining cats and dogs!")
'("cats and dogs")
```

The period in a regular expression will match any single character, so the regular expression above will match any substring that has the string `"cats"` followed somewhere else by the string `"dogs"`.

What if we just want to know if the string contains `"cats"` *or* `"dogs"`? This is where the regular expression *or* operator, which consists of a vertical bar (`|`), comes into play.

```
> (regexp-match #rx"cats|dogs" "Do you like cats?")
'("cats")
> (regexp-match #rx"cats|dogs" "Or do you like dogs?")
'("dogs")
```

The circumflex (`^`) and and dollar sign (`$`) characters are special regular expression markers. The circumflex indicates that the match must start at the beginning of the string or, if a start position is specified, at the start position. Likewise, the dollar sign indicates that the match must extend to the end of the string or the ending position, if specified.

```
> (regexp-match #rx"^[0-9]+" "Is the number 1234 an integer?" 16)
'("34")

> (regexp-match #rx"^[0-9]+" "Is the number 1234 an integer?")
#f

> (regexp-match #rx"^[0-9]+" "987 is an integer!")
'("987")

> (regexp-match #rx"[0-9]+$" "987 is a number?")
#f

> (regexp-match #rx"[0-9]+$" "The number is at the end: 987")
'("987")
```

Table 10-1 provides a summary description of the various regular expression operators. The string "..." in the table represents an arbitrary list of characters.

Table 10-1: Regular Expression Operators

Operator	Description
.	Match any character
*x**	Match *x* zero or more times
x+	Match *x* one or more times
x?	Match *x* zero or one time
x | *y*	Match *x* or *y*
^	Match from start of string
$	Match to end of string
[...]	Define character class
[^...]	Define excluded character class

One thing that's not obvious in the discussion so far is that each letter and digit is in fact a regular expression. A string such as "abc" is actually the concatenation of the letters a, b, and c. Much like multiplication in a mathematical expression such as $3a$, concatenation is implicit in regular expressions. Also like multiplication versus addition, concatenation has a higher precedence than the or (|) operator. This means that an expression like "abc |def" is interpreted as "(abc)|(def)" instead of "ab(c|d)ef" (note the parentheses in these last two strings are just examples of how the regular expression "abc|def" is *interpreted*, but see the following for more on how parentheses play into regular expressions).

Parentheses are used in regular expressions to group subexpressions together and to specify order of evaluation. Let's see how this plays out.

```
> (regexp-match #rx"abc|def" "abcdef")
'("abc")

> (regexp-match #rx"abc|def" "defabc")
'("def")

> (regexp-match #rx"(abc)|(def)" "abcdef")
'("abc" "abc" #f)

> (regexp-match #rx"ab(c|d)ef" "abcdef")
#f

> (regexp-match #rx"ab(c|d)ef" "abcef")
'("abcef" "c")
```

The first two examples return the first part of the string that matches either "abc" or "def".

The third example, using subexpressions, returns three values. The first is the expected match for the overall regular expression. The second value represents the answer to the question: within the first returned value, what is the match for the subexpression "(abc)"? In this case, the value is just the string "(abc)". The third value answers this question: within the first returned value, what is the match for the subexpression "(def)"? In this case there's no match, so it returns #f.

In the fourth example, the match fails because the regular expression is looking for a string with either c or d, but not both. In the last example, the entire string was matched, which is reflected in the first return value, but the second value reflects the fact that only the "c" from the subexpression "(c|d)" was matched.

In our lexical analyzer, we'll want to use subexpressions, but we'll only want to know whether the overall regular expression found a match, and we won't be interested in individual subexpression matches (that is, we're mainly using it to control evaluation). In this case, we'll use a special parentheses syntax, "(?>...)", which indicates that we only want the overall match without bothering to return matched subexpressions (note that ?: works in a similar way to ?>, but ?: allows specifying matching modes, like whether or not the match is case sensitive—see the Racket Documentation for specifics).

```
> (regexp-match #rx"(?>abc)|(?>def)" "abcdef")
'("abc")
```

One interesting variant of regexp-match is regexp-match*. This particular function (although we won't have a need for it in our application) returns the subexpression matches only.

```
> (regexp-match* #rx"(abc)|(def)" "abcdef")
'("abc" "def")
```

Notice that regexp-match only matches "abc", but regexp-match* returns a list of all matches, so both "abc" and "def" are returned. See the Racket Documentation for more on regexp-match*.

> **NOTE** *Racket provides an additional form of regular expressions that conform to the ones used in the Perl programming language. The function used to create regular expressions of this form is called* pregexp. *There's also a literal syntax, similar to the* #rx *form, but starting with* #px *instead. The Perl syntax provides a number of useful extensions, including predefined character classes. Since our needs are fairly simple, we'll stick with the basic syntax outlined above.*

Regular Expressions in TRAC

In TRAC (or any calculator for that matter), we need to identify valid numeric strings (floating-point numbers, to be exact). In addition we'll want to define variables, so that means we'll need to be able to define identifiers. We'll need to specify mathematical operators for addition, subtraction, and so on as well as a judicious set of elementary function names. These items all dictate the use of regular expressions.

For the purposes of our TRAC application, we'll always specify the starting position for the regular expression search, so each regular expression will start with ^. The recognizer for identifiers is defined as follows:

```
(define regex-ident #rx"^[A-Za-z](?>[A-Za-z]|[0-9])*")
```

It should be clear from the information above that this will match any string that starts with a letter and is followed by zero or more letters or digits.

The recognizer for numbers (below) is a bit more involved, but the only new element is the portion with \\.. Since the period (.) is a regular expression that matches any character, it needs to be *escaped* so that it can be treated as a regular character (if a character has a special meaning in regular expressions, escaping is a means of removing, or *escaping*, that special meaning). To avoid having to escape the period, we could also have specified \\. as [.], which might be easier to read in some contexts. The regular expression escape character is the backslash (\), and since it's embedded in a Racket string, it must also be escaped by prefixing it with another slash.

```
(define regex-number #rx"^[0-9]+(?>\\.[0-9]+)?(?>e[+-]?[0-9]+)?")
```

While this is a bit long, it closely mirrors the definition specified by the syntax diagram given earlier in Figure 10-4. Let's review a few test cases.

```
> (regexp-match regex-number "123")
'("123")

> (regexp-match regex-number "a123")
#f
```

```
> (regexp-match regex-number "123.")
'("123")
```

Notice in the last expression that the match didn't include the decimal point, since we specified that a decimal point must be followed by at least one digit. This is in line with the syntax diagram, since the match is up to but doesn't include the decimal point. If the regular expression had ended with `$`, this match would have failed. Notice the following.

```
> (regexp-match regex-number "123.0")
'("123.0")
```

In this case, the entire string is matched. Here are a few more examples.

```
> (regexp-match regex-number "123.456")
'("123.456")

> (regexp-match regex-number "123.456e")
'("123.456")
```

Again, the match didn't include the `e`, since we specified that an `e` must be followed by at least one digit.

```
> (regexp-match regex-number "123.456e23")
'("123.456e23")

> (regexp-match regex-number "123.456e+23")
'("123.456e+23")

> (regexp-match regex-number "123e+23")
'("123e+23")

> (regexp-match regex-number "123e23")
'("123e23")

> (regexp-match regex-number "e23")
#f
```

The definition for arithmetic operators is obvious.

```
(define regex-op #rx"^[-+*/^=]")
```

We'll want to skip over any space characters, so we add this to our toolbox:

```
(define regex-ws #rx"^ *")
```

To make TRAC truly useful, we include the usual transcendental functions.

```
(define regex-fname #rx"^(sin|cos|tan|asin|acos|atan|log|ln|sqrt)")
```

Finally, to facilitate variable assignment, we create a regular expression for keywords. For now, let is our only keyword.

```
(define regex-keyword #rx"^let")
```

The Lexer

With the essential definitions in place, we move on to actually defining the lexical analyzer. Rather than just returning a list of tokens, we're going to supplement each token value with its type. For example, if an identifier is matched, we're going to return a pair: the first element of the pair is the token type, in this case identifier, and the second element is the matched string. This additional information will make the job of the parser a bit easier.

The lexer is (conceptually) fairly simple: it just sequentially tries to match each token type while keeping track of the position of the matched string. If no match is found, the process fails. If a match is found, the token and its position are recorded, and the process repeats at the next position. This continues until the entire input string is consumed.

Another point of interest is that we're using regexp-match-positions as our matching function. This will allow us to easily get the position of the next location once a match has been made.

The TRAC lexical analyzer is a function called tokenize, as given below. The main body of the code is a few lines long (see the cond block ❸); the rest of the code is composed of a few helper functions to manage some of the bookkeeping.

```
(define (tokenize instr)
  (let loop ([i 0])
    (let* ([str-len (string-length instr)]
           [next-pos 0]
         ❶ [start (cdar (regexp-match-positions regex-ws instr i))])

      (define (match-reg regex)
        (let ([v (regexp-match-positions regex instr start)])
          (if (equal? v #f)
              (set! next-pos #f)
              (set! next-pos (cdar v)))
          next-pos))

    ❷ (define (classify type)
        (let ([val (substring instr start next-pos)])
          (if (equal? type 'number)
              (cons type (string->number val))
              (cons type val))))

      (define (at-end)
        (or (= str-len next-pos)
```

```
          (let ([c (string-ref instr next-pos)])
            (not (or (char-numeric? c) (char-alphabetic? c))))))

    (let ([token
         ❸ (cond [(= start str-len)'()]
                ❹ [(and (match-reg regex-keyword) (at-end))
                     (classify 'keyword)]
                  [(and (match-reg regex-fname) (at-end))
                     (classify 'fname)]
                  [(match-reg regex-ident) (classify 'ident)]
                  [(match-reg regex-number) (classify 'number)]
                  [(match-reg regex-op) (classify 'op)]
                ❺ [(equal? #\( (string-ref instr start))
                     (set! next-pos (add1 start))
                     (cons 'lpar "(")]
                ❻ [(equal? #\) (string-ref instr start))
                     (set! next-pos (add1 start))
                     (cons 'rpar ")")]
                  [else #f])])
      (cond [(equal? token '()) '()]
          ❼ [token (cons token (loop next-pos))]
          ❽ [else (error (format "Invalid token at ~a." start))])))))
```

At each iteration of the loop beginning on the second line, the variable `i` has the current starting position within the input string `instr`. After initializing `str-len` and `next-pos`, the function reads past any whitespace ❶. The `match-reg` function executes the regular expression passed to it in `regex` and sets `next-pos` to the next position in the string if there is a match; otherwise it's set to `#f`. If there's a match, `next-pos` is returned; otherwise the function returns `#f`. The `classify` function ❷ merges the token type and the token value into a Racket `cons` cell. If the token is a number, it also converts the string value to the corresponding numeric value. The `at-end` function tests whether the tokenizer is at the end of a keyword or function. A string like `sine` is a valid variable name, but wouldn't be valid as the function name `sin`, so `at-end` allows the tokenizer to differentiate one input type from another.

With these functions available, the actual logic to tokenize the string is fairly straightforward. A check is made ❸ to see if we're at the end of the string, and if so, the empty list is returned. Next is a series of checks ❹ to see whether the text at the current position in the string matches any one of the specified regular expressions; if so, the matching token is packaged up in a `cons` cell by `classify` and returned. If no match is found, the `cond` statement returns `#f`, which results in an error being generated ❽. If the value of `token` is anything other than `#f`, it's added to the returned list ❼. We didn't bother setting up regular expressions for parentheses, since they can be handled easily ❺ ❻.

The order in which the regular expressions are evaluated is important. If `regex-ident` were evaluated before `regex-fname`, a function name like `cos` could mistakenly be interpreted as an ordinary variable name instead of the cosine function (this could be dealt with in the parser, but it's better to offload as much work as possible to the lexical analyzer).

Here's an example of the output:

```
> (tokenize "(x1*cos(45) + 25 *(4e-12 / alpha)^2")
'((lpar . "(")
  (ident . "x1")
  (op . "*")
  (fname . "cos")
  (lpar . "(")
  (number . 45)
  (rpar . ")")
  (op . "+")
  (number . 25)
  (op . "*")
  (lpar . "(")
  (number . 4e-012)
  (op . "/")
  (ident . "alpha")
  (rpar . ")")
  (op . "^")
  (number . 2))
```

The Parser

Our next major TRAC component is the parser. The parser takes the token list from the lexical analyzer and outputs an abstract syntax tree that can be further processed by either an interpreter or a compiler. We first provide a formal definition of our grammar, which will be used as a guide in the construction of the parser.

TRAC Grammar Specification

Computer languages are often specified by a metasyntax (a syntax that describes another syntax) called *extended Backus–Naur form (EBNF)*. You'll notice many similarities between EBNF and regular expressions, but EBNF has more expressive power. EBNF can be used to describe *context-free grammars*, or *CFG* (see "A Few Words About Languages" on page 272), which are out of the reach of regular expressions. (TRAC utilizes a CFG.) This notation will be used to give a formal definition to TRAC. We're going to begin simply, by formally defining what's meant by `digit` (we're actually going to use the lexical analyzer to recognize numbers and identifiers, but for the sake of introducing simple examples of EBNF, we also define them here).

```
digit = "1" | "2" | "3" | "4" | "5" | "6" | "7" | "8" | "9";
```

This is called a *production rule*. As in regular expressions, the vertical bar (|) means *or*. Items in quotation marks (") are called *terminals*, and the identifier digit is called a *nonterminal*. A terminal is a sequence of actual characters (such as what you type on your computer terminal). A nonterminal is a label for a rule, such as digit above. The definition for letter is similar, but we don't show it here, because you can figure it out.

The production for unsigned follows directly from digit:

```
unsigned = digit , { digit };
```

In EBNF, curly brackets { and } function almost exactly like the Kleene star (except that they also allow grouping items together). This means the items within curly brackets can be repeated zero or more times. The comma (,) is the concatenation operator.

With these entities established, we define identifier as follows:

```
identifier = letter , { letter | digit };
```

The production for number is as follows:

```
number = unsigned , [ "."  unsigned ]
       , [ "e",  [ "+" | "-" ] , unsigned ];
```

This production introduces the use of square brackets [and]. Much like the regular expression ?, square brackets enclose optional items.

Function names are defined as follows:

```
fname = "sin" | "cos" | "tan" | "asin" | "acos" | "atan"
      | "log" | "ln";
```

All these productions have regular expression equivalents, so the implementation is managed by the lexer. The parser will implement more complex production rules. Arithmetic expressions typically contain several levels of nested parenthetical expressions; such expressions constitute a context-free grammar. As mentioned in the previous chapter, parsing such expressions exceeds the capability of an FSA (and by extension, regular expressions). Therefore, we now need the expressive power of EBNF to complete our definitions.

With these preliminaries out of the way, we can now give the rest of the definition of the TRAC grammar. Since we only use production names without spaces, commas will be omitted, and therefore concatenation is implicit.

```
statement = "let" identifier "=" expr
          | expr;

expr = term { [ "+" | "-" ] term };

term = neg { [ "*" | "/" ] neg };

neg = "-" neg
    | pow;

pow = factor | factor "^" pow;
```

```
factor = number
       | identifier
       | "(" expr ")"
       | fname  "(" expr ")";
```

These rules are written in such a way that the higher-precedence operators are nested further down. Because of how EBNF is evaluated (example below), this ensures that multiplication and division occurs before addition and subtraction. Likewise, exponentiation occurs ahead of multiplication and division. Note also that the `pow` production is defined recursively, with the recursive call to the right of the operator. This makes exponentiation right-associative, which is how it's normally handled (that is, `a^b^c` is interpreted as `a^(b^c)`, where the rightmost exponentiation is performed first).

Table 10-2 illustrates how the productions are expanded for the expression $a * (1 + b)$.

Table 10-2: Expansion of $a * (1 + b)$

1		*statement*					
2		*expr*					
3		*term*					
4	*neg*	*term-op*	*neg*				
5	*pow*	*term-op*	*neg*				
6	*factor*	*term-op*	*neg*				
7	*identifier*	*term-op*	*neg*				
8	a	*term-op*	*neg*				
9	a	*	*pow*				
10	a	*	*factor*				
11	a	*	(		*expr*		)
12	a	*	(	*term*	*expr-op*	*term*	)
13	a	*	(	*neg*	*expr-op*	*term*	)
14	a	*	(	*pow*	*expr-op*	*term*	)
15	a	*	(	*factor*	*expr-op*	*term*	)
16	a	*	(	*number*	*expr-op*	*term*	)
17	a	*	(	1	*expr-op*	*term*	)
18	a	*	(	1	+	*neg*	)
19	a	*	(	1	+	*pow*	)
20	a	*	(	1	+	*factor*	)
21	a	*	(	1	+	*identifier*	)
22	a	*	(	1	+	b	)

Standard typeface is used to designate terminal tokens, and italics are used to designate nonterminal rules. The notation *expr-op* refers to the expression operators + and -, and *term-op* refers to the term operators * and /. Notice that only the leftmost production is expanded until a terminal value is recognized. Expansion starts with the *statement* rule on row 1. A *statement* can be an *expr*, which in turn can be a *term*; this is reflected on rows 2 and 3.

A *term* can be a *neg* followed by a *term-op* followed by a *neg*. This is shown on row 4. Expansion continues in this fashion until we get to row 7. Notice that our leftmost rule is *identifier*. We now have a terminal, a, that satisfies this rule. The expansion of this rule is shown on row 8. The leftmost rule on this row is *term-op*, which can be expanded to the terminal *. Expansion continues in this way until we have parsed the entire string on row 22.

This grammar is designed in such a way that it's an *LL(1) grammar*. The term LL(1) means that it scans its input (the list of tokens from the lexer) left to right, using a leftmost derivation (as we did in the walk-through above), with a lookahead (lookahead just defines how far ahead we need to look into the list of input tokens) of one symbol (token). This particular type of grammar allows parsers to be constructed in such a way that no backtracking is required to parse the input stream. LL(1) grammars are recognized by *recursive descent parsers* in which each nonterminal production has a procedure (or function) that's responsible for recognizing its portion of the grammar and returns the corresponding portion of the syntax tree (or generating an error if the input is incorrect).

The TRAC Parser

As mentioned in the previous section, TRAC will use a recursive descent parser. A recursive descent parser is mainly a set of mutually recursive functions where there's a function for each grammar rule. There's always a starting function (corresponding to the top-level rule—which is why this is called a top-down parser), which calls other functions as defined by the grammar. The *descent* part of the definition comes about due to the fact that the rules continue to nest down until a terminal (or error) is encountered.

We need a few global variables to keep track of the tokens during the parsing process.

```
(define token-symbol #f)
(define token-value null)
(define token-list '())
```

Next are the predicates used to test various operator types.

```
(define (assign?) (equal? token-value "="))

(define (pow?) (equal? token-value "^"))

(define (neg?) (equal? token-value "-"))

(define (term?) (or
                 (equal? token-value "*")
                 (equal? token-value "/")))

(define (expr?) (or
```

```
                  (equal? token-value "+")
                  (equal? token-value "-")))
```

The following procedure updates the token info whenever the next token value is requested.

```
(define (next-symbol)
  (unless (null? token-list)
    (let ([token (car token-list)])
      (set! token-symbol (car token))
      (set! token-value (cdr token))
      (set! token-list (cdr token-list)))))
```

The accept function tests whether the input token is of the expected type and, if so, reads in the next token and returns #t; otherwise, it returns #f.

```
(define (accept sym)
  (if (equal? token-symbol sym)
      (begin
        (next-symbol)
        #t)
      #f))
```

The expect function tests whether the input token is of the expected type and, if so, reads in the next token and returns #t; otherwise, it generates an error.

```
(define (expect sym)
  (if (accept sym)
      #t
      (if (null? token-list)
          (error "Unexpected end of input.")
          (error (format "Unexpected symbol '~a' in input." token-value)))))
```

The reason we have both accept and expect is that in some cases we need to test for various token types without generating an error. For example, the *factor* rule accepts a number of different token types. We don't want to generate an error if we're testing for a number and the current token is an identifier, because if the number test fails, we still want to test for an identifier, so we use accept. On the other hand, if the expected token *must* be of a particular type, we use the expect function, which generates an error if the current token isn't of the expected type.

We're now able to define the functions that correspond to each grammar production. Even though recursive descent parsers are top-down parsers, we're going to present the code from the bottom up. Since there are fewer dependencies that way, it should be easier to understand. Given that, the first function is factor:

```
(define (factor)
  (let ([val token-value])
```

```
    (cond [(accept 'number) (cons 'number val)]
          [(accept 'ident) (cons 'ident val)]
        ❶ [(accept 'lpar)
           (let ([v (expr)])
             (expect 'rpar)
             v)]
          [(accept 'fname)
           (let ([fname val])
             (expect 'lpar)
             (let ([v (expr)])
               (expect 'rpar)
               (cons 'func-call (cons fname v))))]
          [else (error (format "Invalid token: ~a." token-value))])))
```

Note that we need to save the current token value in `val` (which is set in the first line of `let`). Once `accept` is called and a match is found, the variable `token-value` is set to the value of the next token, which isn't what we need in the return value in the `cond` section of the code. The correspondence between the various `cond` tests and the production for `factor` should be self-evident. As a bit of explanation for the third condition branch ❶, if we look back at our rule for `factor`, we find `"(" expr ")"` as an accepted production. So we see that this portion of the code accepts a left parenthesis, calls `expr` to parse that part of the rule, and then *expects* a right parenthesis (and errors out if that isn't the current token).

For each accepted value, a `cons` cell is created where the first element is a symbol identifying the node type and the second element is the value. The function call portion of the `factor` rule (`fname "(" expr ")"`) wasn't given a name, but we specify `'func-call` here to identify the node type. This pattern of defining functions for rules will be replicated in all the productions, with the end result being the desired parser to construct the syntax tree.

Next up is the code for `pow`:

```
(define (pow)
  (let ([e1 (factor)])
    (if (pow?)
        (begin
          (next-symbol)
        ❶ (let ([e2 (pow)])
            (cons '^ (cons e1 e2))))
        e1)))
```

This is written in such a way to enforce the grammar rule that requires it to be right-associative. This is managed by the recursive call to `pow` ❶. The value returned for `pow` is either just the value returned from `factor` or a new pair (if the symbol `^` is recognized). The first element of this new pair is the character `^` and the second element is another pair, where the first element is the base number (from `e1`) and the second element is the power it's being raised to (from a recursive call to `pow`).

The code for neg (unary minus) is quite simple. If needed, it appends a negation operator to the return value from pow to generate a node for unary minus.

```
(define (neg)
  (if (neg?)
      (begin
        (next-symbol)
        (cons 'neg (neg)))
      (pow)))
```

Multiplication and division are handled by the next function, term. As long as it keeps recognizing other term operators (* or /), it loops, gathering values from neg. Notice how this differs from the code for pow: this code makes term operators left-associative whereas the code for pow makes exponentiation right-associative.

```
(define (term)
  (let ([e1 (neg)])
    (let loop ()
      (when (term?)
        (let ([op (if (equal? token-value "*") '* '/)])
          (next-symbol)
          (let ([e2 (neg)])
            (set! e1 (cons op (cons e1 e2)))))
        (loop)))
    e1))
```

Addition and subtraction are managed by expr. This function works analogously to term.

```
(define (expr)
  (let ([e1 (term)])
    (let loop ()
      (when (expr?)
        (let ([op (if (equal? token-value "+") '+ '-)])
          (next-symbol)
          (let ([e2 (term)])
            (set! e1 (cons op (cons e1 e2)))))
        (loop)))
    e1))
```

Finally, we get to the top level, where most of the work that needs to be done is to set things up to parse the assignment statement.

```
(define (statement)
  (if (equal? token-value "let")
      (begin
        (next-symbol)
        (let ([id token-value])
```

```
        (accept 'ident)
        (if (assign?)
            (begin
              (next-symbol)
              (cons 'assign (cons id (expr))))
            (error "Invalid let statement"))))
    (expr)))
```

The actual parser just has to call tokenize (the lexer) to convert the input string to a list of tokens and kick off the parsing process by calling statement.

```
(define (parse instr)
  (set! token-list (tokenize (string-trim instr)))
  (next-symbol)
  (let ([val (statement)])
    (if (equal? token-list '())
        val
        (error "Syntax error in input."))))
```

Notice that if there's anything left in token-list, an error is generated. Without this, an input that starts with a valid expression, but has some dangling tokens. For example, the following would return a partial result (in this case '(ident . "x")) without alerting the user that the input was invalid.

```
> (parse "x y")
```

Here it goes with a test input expression:

```
> (parse "(x1*cos(45) + 25 *(4e-12 / alpha))^2")
'(pow (+ (* (ident . "x1") func-call "cos" number . 45) * (number . 25) / (
    number . 4e-012) ident . "alpha") number . 2)
```

It seems to work, but it's a bit difficult to decipher what's actually going on with this output. We need a procedure that will take the syntax tree and print it in a way that makes the structure more obvious. So here it is!

```
(define (print-tree ast)
  (let loop ([level 0][node ast])
    (let ([indent (make-string (* 4 level) #\ )]
          [sym (car node)]
          [val (cdr node)])
      (printf indent)
      (define (print-op)
        (printf "Operator: ~a\n"  sym)
        (loop (add1 level) (car val))
        (loop (add1 level) (cdr val)))
      (match sym
        ['number (printf "Number: ~a\n" val)]
        ['ident  (printf "Identifier: ~a\n" val)]
        ['func-call
```

```
        (printf "Function: ~a\n" (car val))
        (loop (add1 level) (cdr val))]
      ['+ (print-op)]
      ['- (print-op)]
      ['* (print-op)]
      ['/ (print-op)]
      ['^ (print-op)]
      ['neg
        (printf "Neg:\n")
        (loop (add1 level) val)]
      ['assign
        (printf "Assign: ~a\n" (car val))
        (loop (add1 level) (cdr val))]
      [_ (printf "Node: ~a?\n" node)]))))
```

It's essentially one big match statement that matches against the node type of the tree. The indentation varies depending on the depth of the node in the tree. This will provide a visual representation of how the child nodes are lined up. With this, we can generate output that is a bit more decipherable.

```
> (define ast (parse "(x1*cos(45) + 25 *(4e-12 / alpha))^2"))
> (print-tree ast)
Operator: ^
    Operator: +
        Operator: *
            Identifier: x1
            Function: cos
                Number: 45
        Operator: *
            Number: 25
            Operator: /
                Number: 4e-012
                Identifier: alpha
    Number: 2
```

The parser creates a syntax tree of an input string, and print-tree prints out a visual representation of the tree. It turns out that print-tree provides a framework with which to build a routine that can reconstruct the input string from the syntax tree. This can be useful for debugging purposes, since it allows us to see whether an output string constructed from the AST corresponds to the input string. We reverse the process by first creating a token list from the syntax tree, and then we create an output string by appending the tokens together.

The biggest issue in creating a tree-to-string conversion function lies in deciding when to add parentheses around an expression. We certainly want to include them when required, but we don't want to include unnecessary parentheses when they aren't required. To facilitate this, we create a function that returns the precedence and associativity of each operator. This is needed to determine whether or not parentheses are required (for example, operators with lower precedence will require parentheses, and if the precedence is the same, the need for parentheses is dictated by the associativity).

```
(struct op (prec assoc))

(define get-prop
  (let ([op-prop
         (hash
          'assign (cons 0 'r)
          '+    (op 10 'l)
          '-    (op 10 'l)
          '*    (op 20 'l)
          '/    (op 20 'l)
          'neg (op 30 'n)
          '^    (op 40 'r)
          'expt    (op 40 'r)
          'number (op 90 'n)
          'ident   (op 90 'n)
          'func-call (op 90 'n))])
    (λ (sym)
      (hash-ref op-prop sym (λ () (op 90 'n))))))
```

If a symbol isn't in the table, the second λ expression returns a default value of (info 90 'n).

With this function at hand, we can produce ast->string:

```
(define (ast->string ast)
  (let ([expr-port (open-output-string)])
    (define (push str)
      (display str expr-port))
    (let loop ([node ast])
      (let* ([sym (car node)]
             [val (cdr node)]
             [prop (get-prop sym)]
             [prec (op-prec prop)]
             [assoc (op-assoc prop)])

        (define (need-paren arg side)
          (let ([arg-prec (op-prec (get-prop (car arg)))])
            (cond [(< arg-prec prec) #t]
                  [(> arg-prec prec) #f]
                  [else (not (equal? assoc side))])))

        (define (push-op)
          (let* ([e1 (car val)]
                 [par1 (need-paren e1 'l)]
                 [e2 (cdr val)]
```

```
                    [par2 (need-paren e2 'r)])
                (when par1 (push "("))
                (loop e1)
                (when par1 (push ")"))
                (push (format " ~a "  sym))
                (when par2 (push "("))
                (loop e2)
                (when par2 (push ")"))))

        (match sym
          ['number (push (number->string val))]
          ['ident (push val)]
          ['func-call
           (push (car val))
           (push "(")
           (loop (cdr val))
           (push ")")]
          ['+ (push-op)]
          ['- (push-op)]
          ['* (push-op)]
          ['/ (push-op)]
          ['^ (push-op)]
          ['neg
           (push "-")
           (let ([paren (need-paren val 'n)])
             (when paren (push "("))
             (loop val)
             (when paren (push ")")))]
          ['assign
           (push (format "let ~a = " (car val)))
           (loop (cdr val))]
          [_ (push (format "Node: ~a" sym))])))
    (get-output-string expr-port)))
```

A local function called push has been defined that adds a token to the output string port (expr-port). One major difference between this code and print-tree is that all the print statements have been changed to push statements. In addition, the function that handles the various operators, push-op (instead of print-op), has been expanded to decide when to include parentheses. Aside from these changes, the structural similarities, starting with the match statement, between ast->string and print-tree should be fairly obvious. So now we can go full circle: input string to abstract syntax tree and back to input string:

```
> (ast->string (parse "(x1*cos(45) - 4 + -25 *(4e-12 / alpha))^2"))
"(x1 * cos(45) - 4 + -25 * (4e-012 / alpha)) ^ 2"
```

TRAC

Once the syntax tree has been created, the rest of the work is smooth sailing. The main remaining components are a dictionary to hold our variable values and the code that actually evaluates our input expressions and produces a numeric value. Before we wrap up, we'll look at a few enhancements, such as adding complex numbers and setting the angular mode.

Adding a Dictionary

Since TRAC has the capability to assign values to variables, we'll need a dictionary to hold the values. We're actually going to create this in the form of a function, where we pass it an action (for example, get to retrieve a value and set to assign a value). This will make it easier to extend its functionality without cluttering up the namespace with additional definitions. This also provides an example of using a single *rest-id* in a lambda expression. A rest-id is a parameter that takes all the arguments supplied to the function in a single list. The args parameter in the code below is the rest-id that accepts a list of arguments. Notice that it's not surrounded by parentheses.

```
(define var
  (let ([vars (make-hash)])
    (λ args
      (match args
        [(list 'set v n) (hash-set! vars v n)]
        [(list 'get v)
         (if (hash-has-key? vars v)
             (hash-ref vars v)
             (error (format "Undefined variable: ~a" v)))]))))
```

Observe that this code actually uses a closure to construct the dictionary (that is, vars, in the form of a hash table). This function returns a function that has the dictionary embedded in it.

With a dictionary to hold variable values in place, we can now define the expression evaluator.

```
(define (eval-ast ast)
  (let loop ([node ast])
  ➊ (let ([sym (car node)]
          [val (cdr node)])

   ➋ (define (eval-op)
       (let ([n1 (loop (car val))]
             [n2 (loop (cdr val))])
         (match sym
           ['+ (+ n1 n2)]
           ['- (- n1 n2)]
           ['* (* n1 n2)]
           ['/ (/ n1 n2)]
```

```
          ['^ (expt n1 n2)]))))

❸ (define (eval-func fname val)
    (match fname
      ["sin" (sin val)]
      ["cos" (cos val)]
      ["tan" (tan val)]
      ["asin" (asin val)]
      ["acos" (acos val)]
      ["atan" (atan val)]
      ["ln" (log val)]
      ["log" (log val 10)]
      ["sqrt" (sqrt val)] ))

❹ (match sym
    ['number val]
    ['ident (var 'get val)]
    ['+ (eval-op)]
    ['- (eval-op)]
    ['* (eval-op)]
    ['/ (eval-op)]
    ['^ (eval-op)]
    ['neg (- (loop val))]
    ['assign (var 'set (car val)
                  (loop (cdr val)))]
  ❺ ['func-call
     (eval-func (car val)
                (loop (cdr val)))]
    [_ (error "Unknown symbol")]))))
```

Notice that it follows a pattern similar to `ast->string` and `print-tree`; the difference is that now, instead of returning or printing a string, it traverses the syntax tree and computes the numerical values of the nodes.

Let's walk through what happens. Given the AST, we extract the parsed symbol (`sym`) and value (`val`) ❶. We then match the symbol ❹ and take the appropriate action. If we're given a literal number, we simply return the value. If we have an identifier, then we extract the value from the dictionary using `(var 'get val)`. An arithmetic operation will result in calling `eval-op` ❷, which first recursively extracts arguments `n1` and `n2`. It then matches the input symbol to determine which operation to perform. A function call ❺ recursively extracts its argument via `(loop (cdr val))` and calls `eval-func` ❸ to actually perform the computation.

We're now in a position to actually perform some calculations.

```
> (eval-ast (parse "let x = 3"))
> (eval-ast (parse "let y = 4"))
> (eval-ast (parse "sqrt(x^2 + y^2)"))
5
```

```
> (eval-ast (parse "x + tan(45 * 3.14159 / 180)"))
3.9999986732059836
```

To keep from having to call parse and eval-ast every time, we need to set up a read-evaluate-print loop (REPL). To do this, we create a start function that kicks off the process and sets up a few predefined variables.

```
(define (start)
  (var 'set "pi" pi)
  (var 'set "e" (exp 1))
  (display "Welcome to TRAC!\n\n")
  (let loop ()
    (display "> ")
      (let ([v (eval-ast (parse (read-line)))])
        (when (number? v) (displayln  v)))
    (loop)))
```

Now we can exercise TRAC in a more natural way.

```
> (start)
Welcome to TRAC!

> let x = 3
> let y = 2+2
> sqrt(x^2+y^2)
5

> tan(45 * pi / 180)
0.9999999999999999
```

A Few Enhancements

We've now established the basic functionality of TRAC, but to make it truly useful, we'll add a few enhancements. One important enhancement is to have it fail gracefully if the user makes an input error. It might also be nice to provide advanced users the ability to work with complex numbers. We'll explore these topics and more in the sections that follow.

Exception Handling

As it stands, TRAC is quite fragile. The slightest misstep will cause it to fail:

```
> let x=3
> let y=4
> sqrt(x^2 + y^2
. . Unexpected end of input
```

It should be more forgiving of erroneous input (we're human, after all). To alleviate this situation, we leverage Racket's *exception handling* capability.

When an error occurs in executing Racket code, an exception is raised. An exception will either have a type of `exn` or one of its subtypes. The exceptions raised by `error` have a type of `exn:fail`. To trap such errors, one wraps the code in a `with-handlers` form. A modified version of `start` that uses `with-handlers` is given here.

```
(define (start)
  (var 'set "pi" pi)
  (var 'set "e" (exp 1))
  (display "Welcome to TRAC!\n\n")
  (let loop ()
    (display "> ")
    (with-handlers ([exn:fail? (λ (e) (displayln "An error occured"))])
      (let ([v (eval-ast (parse (read-line)))])
        (when (number? v) (displayln v))))
    (loop)))
```

The `with-handlers` form can trap any number of different types of error. In this case, we use the `exn:fail?` predicate to trap generated `exn:fail` errors generated by the `error` form. Each trapped error type has a corresponding function to manage the trapped error.

Here we use a lambda expression to generate the somewhat uninformative `"An error occurred."` message. Evaluating the expression with the missing right parenthesis now produces the following outcome.

```
> sqrt(x^2 + y^2
An error occurred!
>
```

Observe that this time, even though an error has occurred, the > prompt appears, indicating that the program is still running. The user now has an opportunity to re-enter the expression and continue working.

Suppose we want to provide a more informative error message, like the one provided by Racket. The `e` parameter handed to the exception handling function is an `exn` structure. This structure has a `message` field that contains the actual text string of the raised error. So to print the text of the error message, we need to modify the lambda function to read as follows:

```
(λ (e) (displayln (exn-message e)))
```

With this change in place, a session with an erroneous entry would proceed as follows:

```
> (start)
Welcome to TRAC!

> let x=3
> let y=4
> sqrt(x^2 + y^2
Unexpected end of input.
```

```
> sqrt(x^2 + y^2)
5
>
```

Notice that evaluating an expression such as `sqrt(-1)` will produce the complex number `0+1i`. This may be confusing to users not familiar with complex numbers. In this case, it may be preferable to raise an error instead of returning a result. To accommodate this, the `start` procedure could be modified as follows:

```
(define (start)
  (reset)
  (let loop ()
    (display "> ")
    (with-handlers ([exn:fail? (λ (e) (displayln (exn-message e)))])
      (let ([v (eval-ast (parse (read-line)))])
        (when (number? v)
          (if (not (real? v))
              (error "Result undefined.")
              (displayln v)))))
    (loop)))
```

With this change in place, evaluating an expression that returns a complex number would produce the following result:

```
> sqrt(-2)
Result undefined.
```

Complex Numbers

In the previous section, we mentioned throwing an exception if a calculation produces a complex number. If users *are* familiar with complex numbers, the lexer could be modified to accept complex numbers, in which case the original `start` procedure could be kept in place. It's not extremely difficult to modify TRAC's lexical analyzer such that it works with complex numbers. One might be tempted to create a regular expression that recognizes a complex number such as `1+2i`. That would be a big mistake. If one evaluates an expression such as `2*1+2i`, the expected result is `2+2i` since multiplication has a higher precedence than addition. If the lexer returns the entire expression as a number, the parser will treat the expression `2*1+2i` as `2*(1+2i)`, which will give the result `2+4i`.

The actual solution is quite simple. Instead of recognizing the entire complex number, we only recognize the imaginary part. That is, the regular expression for a number becomes as follows:

```
(define regex-number #rx"^[0-9]+(?>\\.[0-9]+)?(?>e[+-]?[0-9]+)?i?")
```

Notice that the only change in the expression is the inclusion of `i?` at the end, which means we accept an optional `i` at the end of a numeric input.

In addition, we make a small modification to `classify` (which is embedded in `tokenize`) to handle imaginary numbers.

```
(define (tokenize instr)
        ⋮
      (define (classify type)
        (let ([val (substring instr start next-pos)])
          (if (equal? type 'number)
              (cons type
                    (if (equal? #\i (string-ref val (sub1 (string-length val))
    ))
                        (string->number (string-append "0+" val))
                        (string->number val)))
              (cons type val))))
        ⋮
```

With these changes in place, we can compute the following in TRAC:

```
> 1i
0+1i

> 1i^2
-1

> 2*1+2i
2+2i

> 2*(1+2i)
2+4i
```

Mode, Reset, and Help Commands

Most calculators allow the user to compute trigonometric functions using either degrees or radians. We'd be remiss to omit this capability from TRAC. This will require a global variable to contain the trigonometric mode:

```
(define RADIANS 1)
(define DEGREES (/ pi 180))
(define trig-mode RADIANS)
```

TRAC currently handles numeric entries exactly as Racket would. That is, if an exact value is divided by an exact value, a fraction results. For example entering `2/4` would return a result of `1/2`. This is typically not what's expected for run-of-the-mill calculations. We'll thus modify TRAC to give the user the option to treat all entries as floating-point numbers or to retain fractional entries. To enable this, we'll use a global variable to maintain the numeric mode.

```
(define FRAC 1)
(define FLOAT 2)
```

```
(define num-mode FLOAT)
```

It would also be nice to allow the user to reset TRAC to its default start-up state, so TRAC is given a new keyword called reset, which requires the following change to regex-keyword.

```
(define regex-keyword #rx"^(let|reset|\\?)")
```

The question mark at the end will allow TRAC to have a mini-help system, which is accessed by entering ? on the command line (more on this shortly).

Entering reset will result in clearing previous entries in the TRAC dictionary and priming it with the default values. These actions are bundled up into a reset procedure:

```
(define (reset)
  (var 'set "pi" pi)
  (var 'set "e" (exp 1))

  (var 'set "Rad" RADIANS)
  (var 'set "Deg" DEGREES)
  (set! trig-mode RADIANS)

  (var 'set "Frac" FRAC)
  (var 'set "Float" FLOAT)
  (set! num-mode FLOAT)

  (displayln "** Welcome to TRAC! **\n")
  (displayln "  Modes: Rad, Float")
  (displayln "  Enter ? for help.\n")
  )
```

The start procedure then becomes as follows:

```
(define (start)
  (reset)
  (let loop ()
    (display "> ")
    (with-handlers ([exn:fail? (λ (e) (displayln (exn-message e)))])
      (let ([v (eval-ast (parse (read-line)))])
        (when (number? v) (displayln v))))
    (loop)))
```

To accommodate the new reset and ? keywords, the statement portion of the parser is updated as follows:

```
(define (statement)
  (cond [(equal? token-value "let")
         (next-symbol)
         (let ([id token-value])
```

```
        (accept 'ident)
        (if (assign?)
            (begin
              (next-symbol)
              (cons 'assign (cons id (expr))))
            (error "Invalid let statement")))]
      [(equal? token-value "reset") (cons 'reset null)]
      [(equal? token-value "?") (cons 'help null)]
      [else (expr)]))
```

If reset or ? is entered for input, the function returns immediately without drilling down into the parser so that the expression evaluator can handle these commands directly.

Of course we still need to modify the trigonometric functions to work properly depending on the current mode. The handling of numeric entries will also need to be adjusted to ensure that they honor the current numeric mode. Here's the tweaked version of ast-eval.

```
(define (eval-ast ast)
  (let loop ([node ast])
    (let ([sym (car node)]
          [val (cdr node)])

      (define (eval-op)
        (let ([n1 (loop (car val))]
              [n2 (loop (cdr val))])
          (match sym
            ['+ (+ n1 n2)]
            ['- (- n1 n2)]
            ['* (* n1 n2)]
            ['/ (/ n1 n2)]
            ['^ (expt n1 n2)])))

      (define (eval-func fname val)
        (match fname
       ❶ ["sin" (sin (* val trig-mode))]
          ["cos" (cos (* val trig-mode))]
          ["tan" (tan (* val trig-mode))]
          ["asin" (/ (asin val) trig-mode)]
          ["acos" (/ (acos val) trig-mode)]
          ["atan" (/ (atan val) trig-mode)]
          ["ln" (log val)]
          ["log" (log val 10)]
          ["sqrt" (sqrt val)] ))

      (match sym
        ['number
      ❷ (if (and (= num-mode FLOAT) (exact? val))
```

```
            (exact->inexact val)
            val)]
       ['ident (var 'get val)]
       ['+ (eval-op)]
       ['- (eval-op)]
       ['* (eval-op)]
       ['/ (eval-op)]
       ['^ (eval-op)]
       ['neg (- (loop val))]
    ❸ ['reset (reset)]
    ❹ ['help (print-help)]
       ['assign
        (var 'set (car val)
             (let ([n (loop (cdr val))])
             ❺ (cond [(equal? (car val) "TrigMode")
                        (if (or (= n RADIANS) (= n DEGREES))
                            (begin
                              (set! trig-mode n)
                              (printf "TrigMode set to ~a.\n\n" (if (= n
RADIANS) "Rad" "Deg")))
                            (error "Invalid TrigMode."))]
                   ❻ [(equal? (car val) "NumMode")
                        (if (or (= n FRAC) (= n FLOAT))
                            (begin
                              (set! num-mode n)
                              (printf "NumMode set to ~a.\n\n" (if (= n FRAC) "
Frac" "Float")))
                            (error "Invalid NumMode."))]
                       [else n])))]
       ['func-call
        (eval-func (car val)
                   (loop  (cdr val)))]
       [_ (error "Unknown symbol")]))))
```

The actual changes to the trigonometric functions are minor: just a multiplication or division by `mode` does the trick (observe how `trig-mode` is handled ❶). Code is also added to properly convert exact values to inexact when the mode is set to `FLOAT` ❷. Most of the remaining changes involve modifying the assignment statement to trap changes to `TrigMode` ❺ and `NumMode` ❻ to ensure that they can only be assigned proper values. Note the additions for `reset` ❸ and `help` ❹. The `print-help` procedure is provided here:

```
(define (print-help)
  (let ([help (list
               (format "Current TrigMode: ~a"
                       (if (= trig-mode RADIANS) "Rad" "Deg"))
               "To change TrigMode: to radians type:"
               "   let TrigMode = Rad"
```

```
                "To change TrigMode to degrees type:"
                "   let TrigMode = Deg"
                ""
                (format "Current NumMode: ~a"
                        (if (= num-mode FLOAT) "Float" "Frac"))
                "To change NumMode to float type:"
                "   let NumMode: = Float"
                "To change NumMode: to fraction type:"
                "   let NumMode: = Frac"
                ""
                "To reset TRAC to defaults type:"
                "   reset")])
    (let loop([h help])
      (unless (equal? h '())
        (printf "~a\n" (car h))
        (loop (cdr h)))))
  (newline))
```

Here's a session illustrating the new functionality.

```
> (start)
** Welcome to TRAC! **

  Modes: Rad, Float
  Enter ? for help.

> tan(45)
1.6197751905438615
> let TrigMode=Deg
TrigMode set to Deg.

> tan(45)
0.9999999999999999
> atan(1)
45.0
> let TrigMode=45
Invalid TrigMode.
> let TrigMode=Rad
TrigMode set to Rad.

> cos(pi)
-1.0
> 2/4
0.5
> let NumMode=Frac
NumMode set to Frac.

> 2/4
```

```
1/2
> reset
** Welcome to TRAC! **

  Modes: Rad, Float
  Enter ? for help.

> ?
Current TrigMode: Rad
To change TrigMode: to radians type:
   let TrigMode = Rad
To change TrigMode to degrees type:
   let TrigMode = Deg

Current NumMode: Float
To change NumMode to float type:
   let NumMode: = Float
To change NumMode: to fraction type:
   let NumMode: = Frac

To reset TRAC to defaults type:
   reset

>
```

Pretty cool, eh?

Making Sure TRAC Works Properly

Given the nature of this application, it would be nice to have some degree of comfort that it's performing the calculations properly. If you were using this to calculate the landing trajectory of a spaceship to the moon, it wouldn't do to have it return a calculation that results in the spaceship flying out into empty space instead.

Of course one could sit down and manually enter a large number of test equations into TRAC and verify the results by entering the same equations on some other calculator and seeing if the results are the same. This clearly wouldn't be much fun (or very efficient, for that matter). No, we want an automated process where we can have the computer do all the work. The approach we're going to take is to build a procedure that will generate a random Racket expression. This expression can be evaluated using the Racket `eval` function to get a numeric value. In addition we'll need a function that converts the Racket expression into a TRAC expression string. We can evaluate the TRAC expression to see if it returns the same value. We can then have the computer repeat this process thousands of times to make sure we don't produce any mismatches.

Here's the code for the random Racket expression generator.

```
(define ops
  (vector
   (cons '+ 2)
   (cons '- 1) ; unary minus
   (cons '- 2) ; subtraction
   (cons '* 2)
   (cons '/ 2)
   (cons 'expt 2)
   (cons 'sin 1)
   (cons 'cos 1)
   (cons 'tan 1)
   (cons 'asin 1)
   (cons 'acos 1)
   (cons 'atan 1)
   (cons 'sqrt 1)
   (cons 'log 1) ; natural log
   (cons 'log 2) ; base n log
   ))

(define (gen-racket)
  (let ([num-ops (vector-length ops)])
    (let loop ([d (random 1 5)])
      (if (= d 0)
          (exact->inexact (* (random) 1000000000))
          (let* ([op (vector-ref ops (random num-ops))]
                 [sym (car op)]
                 [args (cdr op)]
                 [next-d (sub1 d)])
            (if (= args 1)
                (list sym (loop next-d))
                (if (equal? sym 'log)
                    (list sym (loop next-d) 10)
                    (list sym (loop next-d) (loop next-d)))))))))
```

An operator from the `ops` vector is randomly selected by the `gen-racket` function. Values in `ops` include both the operator symbol and the number of arguments it's expecting (this is called its *arity*). Notice that both `log` and minus (`-`) have two different arities. The function call `log(x)` (base-10 logarithm) in TRAC is the same as `(log x 10)` in Racket. Then `gen-racket` will build an expression containing from one to five random operations or functions with random floating-point numeric arguments. The result is an actual Racket expression instead of an AST, where its arguments and functions are populated with random values.

Here's a look at some of the expressions that `gen-racket` produces.

```
> (gen-racket)
'(* (cos 25563340.24229431) (cos 112137357.31425005))
```

```
> (gen-racket)
'(log 502944961.7985059 10)

> (gen-racket)
'(sqrt (tan (expt (sqrt 721196577.8863264) (+ 739078577.777451
    744205482.2563056))))
```

Most of the work involves converting the Racket expressions into TRAC expressions.

```
(define (racket->trac expr)
  (let ([out-port (open-output-string)])
    (define (push str)
      (display str out-port))
    (let loop ([node expr])
      (if (number? node)
          (push (number->string node))
          (let* ([sym (car node)]
                 [sym (cond [(equal? sym 'expt) '^]
                            [(equal? sym 'log)
                             (if (= (length node) 2) 'ln 'log)]
                            [(equal? sym '-)
                             (if (= (length node) 2) 'neg '-)]
                            [else sym])]
                 [prop (get-prop sym)]
                 [prec (op-prec prop)]
                 [assoc (op-assoc prop)])

            (define (need-paren arg side)
              (if (not (list? arg))
                  #f
                  (let ([arg-prec (op-prec (get-prop (car arg)))])
                    (cond [(< arg-prec prec) #t]
                          [(> arg-prec prec) #f]
                          [else (not (equal? assoc side))]))))

            (define (push-op)
              (let* ([e1 (second node)]
                     [e2 (third node)]
                     [par1 (need-paren e1 'l)]
                     [par2 (need-paren e2 'r)])
                (when par1 (push "("))
                (loop e1)
                (when par1 (push ")"))
                (push (format " ~a "  sym))
                (when par2 (push "("))
                (loop e2)
                (when par2 (push ")"))))
```

```
      (define (push-neg)
        (let* ([e (second node)]
               [paren (need-paren e 'n)])
          (push "-")
          (when paren (push "("))
          (loop e)
          (when paren (push ")"))))

      (define (push-func)
        (push (format "~a"  sym))
        (push "(")
        (loop (second node))
        (push ")"))

      (match sym
        ['+ (push-op)]
        ['- (push-op)]
        ['* (push-op)]
        ['/ (push-op)]
        ['^ (push-op)]
        ['neg  (push-neg)]
        ['sin  (push-func)]
        ['cos  (push-func)]
        ['tan  (push-func)]
        ['asin (push-func)]
        ['acos (push-func)]
        ['atan (push-func)]
        ['ln   (push-func)]
        ['log  (push-func)]
        ['sqrt (push-func)])))
  (get-output-string out-port))))
```

This is largely an adaptation of the ast->string function, but using the randomly generated Racket expressions created by gen-racket as input instead of the TRAC syntax tree. We've had to make some accommodations to account for the multiple arities of - and log. We also match against the literal function symbols. Aside from these considerations, the code should closely mirror that of ast->string. Here are a few samples of its output.

```
> (racket->trac (gen-racket))

"asin(tan(944670433.0 - 858658023.0 + (918652763.0 + 285573780.0)))"
> (racket->trac (gen-racket))
"sin(atan(364076270.0)) / sqrt(ln(536830818.0))"

> (racket->trac (gen-racket))
"atan(978003385.0)"
```

The basic idea is to automate the following process:

```
> (define r (gen-racket))
> r
'(+ (cos (atan 142163217.6660815)) (log (cos 528420918.36769867)))

> (define v1 (eval r))
> v1
-1.021485300993499

> (define v2 (eval-ast (parse (racket->trac r))))
> v2
-1.021485300993499

> (= v1 v2)
#t
```

So here's our test bench:

```
(define (test n)
  (for ([in-range n])
    (let* ([expr (gen-racket)]
           [v1 (eval expr)]
           [v2 (eval-ast (parse (racket->trac expr)))]
           [delta (magnitude (- v1 v2))])
      (when (> delta 0)
        (displayln "Mismatch:")
        (printf "Racket: ~a\n" expr)
        (printf "TRAC: ~a\n" (racket->trac expr))
        (printf "v1: ~a, v2: ~a, delta: ~a\n\n" v1 v2 delta)))))
```

The result of a computation could potentially result in a complex number (for example, `(sqrt -1)`), so we use `magnitude` to get the absolute value size of the difference between the values.

And here's the output from an initial test run, which in fact indicated that the TRAC evaluation routine wasn't always producing the correct results.

```
> (test 10)
Mismatch:
Racket: (atan (atan (expt 137194961.20152807 513552901.52574974)))
TRAC: atan(atan(137194961.20152807 ^ 513552901.52574974))
v1: 1.0038848218538872, v2: 0.2553534898896325, delta: 0.7485313319642546

Mismatch:
Racket: (- (log (expt (+ 67463417.07939068 342883686.1438599) (sin
     521439863.24302197))) (sqrt (+ (atan 402359159.5913063) (acos
     213010305.84288383))))
```

```
TRAC: ln((67463417.07939068 + 342883686.1438599) ^ sin(521439863.24302197)) -
    sqrt(atan(402359159.5913063) + acos(213010305.84288383))
v1: -23.07001808516913-3.029949988703483i, v2:
    16.55567328478171+0.11164266488631025i, delta: 39.75003171002113
```

The common thread in all the mismatches was the exponentiation operator ^ (mapped from Racket's `expt` function), which was inadvertently defined with the division operator in `eval-ast` (the `eval-ast` code given above is correct, but you can introduce the same error if you want to test this). Once the correction was made, another test run produced the following result.

```
> (test 100000)
>
```

In this case, no news is *good* news.

Making an Executable

There's really no need to have TRAC dependent on the DrRacket environment. Only a few additional steps are required to create an executable file that can be launched without starting DrRacket. The first step is to simply add the `(start)` command to the last line of the definitions file (see below) so that the program starts executing immediately when launched.

```
(start)
```

Racket supports three different types of executables:

Launcher This type of executable will execute the current version of your `.rkt` source file, so it will include the path to the source file in the executable. This will allow your executable to immediately reflect any enhancements to your program. The downside is that you can't move the source file elsewhere or easily share the executable with someone else.

Standalone This version embeds the source file in the executable, so there's no problem moving it to another location on your machine. A standalone executable still depends on the installed Racket DLLs, so it may not work properly if moved to a different machine.

Distribution archive A distribution archive bundles all needed files into an install file. The install file can be used to install TRAC on another machine as long as the destination machine uses the same operating system as the one the archive was created on.

Before you create an executable, it's recommended that debugging be turned off. This can be done by going to the **Choose Language...** dialog (from the Language option on the main menu) and pressing the **Show Details** button. This will open a panel where you should select **No Debugging**. Once this is done, go to the Racket main menu, and from there select **Create Executable . . .** . In the dialog box, you may select which of the three different types of executables you want to create. It's even possible to select a custom icon to give TRAC that personal touch.

Figure 10-7 is a screenshot of TRAC running on our machine.

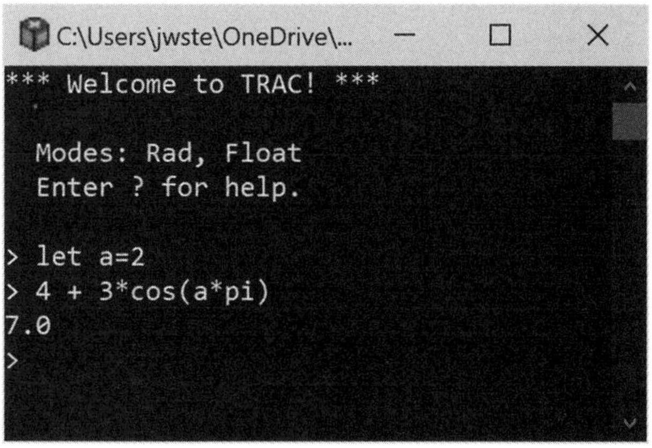

```
C:\Users\jwste\OneDrive\...
*** Welcome to TRAC! ***

  Modes: Rad, Float
  Enter ? for help.

> let a=2
> 4 + 3*cos(a*pi)
7.0
>
```

Figure 10-7: TRAC in action

Summary

In this chapter, we leveraged our knowledge of abstract computing machines and various automata (introduced in the previous chapter) to build an interactive command line expression calculator. Along the way we learned about lexers (and using regular expressions to construct them), parsers (which construct abstract syntax trees), and interpreters. We used EBNF (extended Backus–Naur form) to specify our calculator grammar. Once we had our basic calculator built, we enhanced it with additional capabilities, such as handling complex numbers and hand degrees or radians. Just to be sure our calculator doesn't give us bogus numbers, we built a simple test bench to make certain our code was robust.

Well, that just about concludes our Racket journey for now. But we've only scratched the tip of the iceberg. There's much more capability that we haven't even hinted at. We encourage you to further explore Racket on your own via the Racket website and other available literature. Happy learning!

A

NUMBER BASES

The word *digit* derives from the Latin word *digitius*, meaning finger or toe. Of course we take this as the origin of our counting digits zero through nine, which we can match with our fingers (or toes). These digits form the basis of our base-10 or decimal number system. The positional number system commonly in use today is called the *Hindu–Arabic numeral system*. The works of Muhammad ibn Mūsā al-Khwārizmī (for example *On the Calculation with Hindu Numerals*, c. 825) were influential in the introduction of this system. The innovation was using the position of each digit to represent which power of 10 to multiply the digit by. This made calculations much simpler than using other systems, such as Roman numerals.

The decimal number system assigns to each digit position a power of 10, starting with the rightmost digit in a string of digits. The meaning of a string of digits is then derived by multiplying each digit by the corresponding power of 10, as seen in Figure A-1.

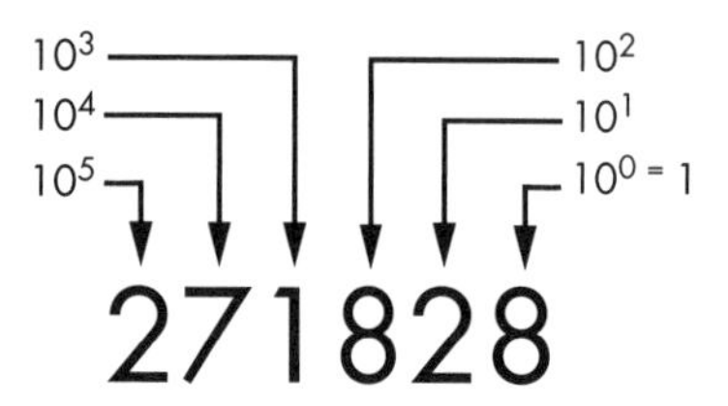

Figure A-1: Base-10 positional values

This is taken to mean the following:

$$271828 = 2 \cdot 10^5 + 7 \cdot 10^4 + 1 \cdot 10^3 + 8 \cdot 10^2 + 2 \cdot 10^1 + 8 \cdot 10^0$$

A concise way to represent this value is as follows:

$$\sum_{i=0}^{5} d_i 10^i$$

Here $d_0 = 8$, $d_1 = 2$, ..., $d_5 = 2$—the digits from least to most significant.

The decimal system isn't the only possible number system. Any integer greater than 1 can be used as the base of a number system. If you were a citizen of a universe where people only had eight fingers (see Figure A-2), you'd probably use the octal (or base-8) number system.

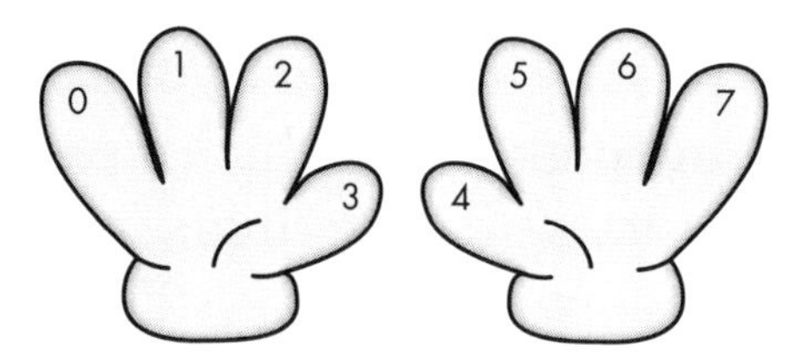

Figure A-2: The octal digits

Numbers in bases other than 10 are normally printed with the base as a subscript of the number. A number in the octal system would be printed as 1234_8. We can convert this to decimal as follows:

$$\begin{aligned} 1234_8 &= 1 \cdot 8^3 + 2 \cdot 8^2 + 3 \cdot 8^1 + 4 \cdot 8^0 \\ &= 1 \cdot 512 + 2 \cdot 64 + 3 \cdot 8 + 4 \cdot 1 \\ &= 668 \end{aligned}$$

As is well known, computers internally work strictly on the binary (or base-2) number system. Three binary digits are capable of representing the numbers from 0 to 7, which, as we've seen, form the basis of the octal number system. Four binary digits can represent the numbers from 0 to 15 (the base-16 digits). Since the numbers 10 through 15 take more than one decimal digit to represent, the letters A through F are used instead. That is,

A=10, B=11, etc. This is known as the hexadecimal number system. For example, the number $FACE_{16} = 64206$, as shown below.

$$\begin{aligned} FACE_{16} &= F \cdot 16^3 + A \cdot 16^2 + C \cdot 16^1 + E \cdot 16^0 \\ &= 15 \cdot 16^3 + 10 \cdot 16^2 + 12 \cdot 16^1 + 14 \cdot 16^0 \\ &= 15 \cdot 4096 + 10 \cdot 256 + 12 \cdot 16 + 14 \cdot 1 \\ &= 64206 \end{aligned}$$

Two hexadecimal digits (or eight binary digits) constitute what's called a byte. Bytes aren't usually used as a number basis, but they're a common unit used to designate the size of computer memory.

Here's some Racket code that takes a positive decimal integer and returns a list consisting of the digits that form the binary representation of the number.

```
(define (decimal->bin n)
  (let loop ([n n] [l '()])
    (if (zero? n) l
        (let-values ([(n d) (quotient/remainder n 2)])
          (loop n (cons d l))))))
```

This works by extracting the least significant (rightmost) binary digit and reducing the value of *n* by dividing by 2 to get to the next digit.

```
> (decimal->bin 15)
'(1 1 1 1)
> (decimal->bin 10)
'(1 0 1 0)
> (decimal->bin 64206)
'(1 1 1 1 1 0 1 0 1 1 0 0 1 1 1 0)
```

Racket also provides the ~r function, which takes a base-10 value and outputs a formatted string in another base.

```
> (~r 64206 #:base 2)
"1111101011001110"

> (~r 64206 #:base 8)
"175316"

> (~r 64206 #:base 16)
"face"

> (~r  170 #:base 2 #:min-width 12  #:pad-string "0" )
"000010101010"
```

B

SPECIAL SYMBOLS

The Racket language supports the usage of Unicode characters in both strings and identifiers. These symbols may be entered directly into DrRacket by typing the codes shown in Tables B-1, B-2, B-3, and B-4. Once the code is entered, hold down the ALT key immediately followed by the \ key (this works in Windows and Linux, but for macOS use CTL-\ instead).

Table B-1: Standard Greek Letters

Symbol	Code	Symbol	Code
α	\alpha	Ξ	\Xi
β	\beta	ξ	\xi
Γ	\Gamma	Π	\Pi
γ	\gamma	π	\pi
Δ	\Delta	ρ	\rho
δ	\delta	Σ	\Sigma
ϵ	\epsilon	σ	\sigma
ζ	\zeta	τ	\tau
η	\eta	Υ	\Upsilon
Θ	\Theta	υ	\upsilon
θ	\theta	Φ	\Phi
ι	\iota	ϕ	\phi
κ	\kappa	χ	\chi
λ	\lambda	Ψ	\Psi
Λ	\Lambda	ψ	\psi
μ	\mu	Ω	\Omega
ν	\nu	ω	\omega

Table B-2: Greek Variants

Symbol	Code
ε	\varepsilon
φ	\varphi
ϖ	\varpi
ϱ	\varrho
ς	\varsigma
ϑ	\vartheta

Table B-3: Other Symbols, Part One

Symbol	Code	Symbol	Code
$\Uparrow$	\Uparrow	$\dagger$	\dagger
$\uparrow$	\uparrow	$\bullet$	\bullet
$\Downarrow$	\Downarrow	$\ddagger$	\ddagger
$\downarrow$	\downarrow	$\wr$	\wr
$\Leftarrow$	\Leftarrow	$\subseteq$	\subseteq
$\leftarrow$	\leftarrow	$\supseteq$	\supseteq
$\Rightarrow$	\Rightarrow	$\subset$	\subset
$\rightarrow$	\rightarrow	$\supset$	\supset
$\nwarrow$	\nwarrow	$\in$	\in
$\swarrow$	\swarrow	$\ni$	\ni
$\searrow$	\searrow	$\notin$	\notin
$\nearrow$	\nearrow	$\neq$	\neq
$\Updownarrow$	\Updownarrow	$\doteq$	\doteq
$\updownarrow$	\updownarrow	$\leq$	\leq
$\Leftrightarrow$	\Leftrightarrow	$\geq$	\geq
$\leftrightarrow$	\leftrightarrow	$\equiv$	\equiv
$\mapsto$	\mapsto	$\sim$	\sim
$\leadsto$	\leadsto	$\cong$	\cong
$\aleph$	\aleph	$\approx$	\approx
$\prime$	\prime	$\propto$	\propto
$\emptyset$	\emptyset	$\models$	\models
∇	\nabla	$\prec$	\prec
$\triangle$	\triangle	$\succ$	\succ
$\neg$	\neg	$\bot$	\bot
$\forall$	\forall	$\top$	\top
$\exists$	\exists	$\vdash$	\vdash
∞	\infty	$\dashv$	\dashv
$\circ$	\circ	$\simeq$	\simeq
$\pm$	\pm	$\ll$	\ll
$\mp$	\mp	$\gg$	\gg
$\cup$	\cup	$\asymp$	\asymp
$\cap$	\cap	$\parallel$	\parallel
$\diamond$	\diamond	$\perp$	\perp

Table B-4: Other Symbols, Part Two

Symbol	Code	Symbol	Code
$\bigtriangleup$	\bigtriangleup	$\bowtie$	\bowtie
$\bigtriangledown$	\bigtriangledown	$\cdot$	\cdot
$\times$	\times	$\sum$	\sum
$\div$	\div	$\prod$	\prod
$\oplus$	\oplus	$\coprod$	\coprod
$\ominus$	\ominus	$\int$	\int
$\otimes$	\otimes	$\oint$	\oint
$\oslash$	\oslash	$\surd$	\sqrt
$\odot$	\odot	☺	\smiley
$\vee$	\vee	☻	\blacksmiley
$\wedge$	\wedge	☹	\frownie
$\diamondsuit$	\diamondsuit	§	\S
$\spadesuit$	\spadesuit	$\vdots$	\vdots
$\clubsuit$	\clubsuit	$\ddots$	\ddots
$\heartsuit$	\heartsuit	$\cdots$	\cdots
$\sharp$	\sharp	$\ldots$	\hdots
$\flat$	\flat	$\langle$	\langle
$\natural$	\natural	$\rangle$	\rangle
$\star$	\star		

BIBLIOGRAPHY

[1] Bridge and torch problem. *https://en.wikipedia.org/wiki/Bridge_and_torch_problem*. Also known as The Midnight Train and Dangerous crossing.

[2] Harold Ableson, Gerald Jay Sussman, and Julie Sussman. *Structure and Interpretation of Computer Programs*. MIT Press, Cambridge, MA, 1993.

[3] W. W. Rouse Ball and H. S. M. Coxeter. *Mathematical Recreations and Essays*. Dover Publications, Inc, New York, NY, 1974.

[4] Norman L. Biggs. *Discrete Mathematics*. Clarendon Press, New York, NY, revised edition, 1989.

[5] William F. Clocksin and Christopher S. Mellish. *Programming in Prolog*. Springer-Verlag, Berlin, fifth edition, 2003.

[6] John H. Conway and Richard K. Guy. *The Book of Numbers*. Copernicus, New York, NY, 1996.

[7] Sarah Flannery. *In Code*. Workman Publishing, New York, NY, 2001.

[8] George Gamow. *One Two Three . . . Infinity*. Viking Press, New York, NY, 1947.

[9] George T. Heineman, Gary Pollice, and Stanley Selkow. *Algorithms in a Nutshell*. O'Reilly, Sebastopol, CA, 2009.

[10] John E. Hopcroft and Jeffrey D. Ullman. *Introduction to Automata Theory, Languages, and Computation*. Addison-Wesley Publishing Company, Reading, MA, 1979.

[11] William J. LeVeque. *Fundamentals of Number Theory*. Addison-Wesley Publishing Company, Reading, MA, 1977.

INDEX

D

E

F

G

H

I

K

L

M

N

X

Z

COLOPHON

The fonts used in *Racket Programming The Fun Way* are New Baskerville, Futura, The Sans Mono Condensed and Dogma. The book was typeset with LaTeX 2_{ε} package `nostarch` by Boris Veytsman *(2008/06/06 v1.3 Typesetting books for No Starch Press).*

The book was printed and bound by Sheridan Books, Inc. in Chelsea, Michigan. The paper is 60# Finch Offset, which is certified by the Forest Stewardship Council (FSC).

The book uses a layflat binding, in which the pages are bound together with a cold-set, flexible glue and the first and last pages of the resulting book block are attached to the cover. The cover is not actually glued to the book's spine, and when open, the book lies flat and the spine doesn't crack.

RESOURCES

Visit *https://nostarch.com/racket-programming-fun-way/* for errata and more information.